Rafael A. Mergener
Antonio C. Alves
Angelo M. Massignam

Effect of plant density on maize cultivation

Rafael A. Mergener
Antonio C. Alves
Angelo M. Massignam

Effect of plant density on maize cultivation

Morphophysiology of an open-pollinated maize variety

ScienciaScripts

Imprint

Cover image: www.ingimage.com

This book is a translation from the original published under ISBN 978-613-9-68091-7.

Publisher:
Sciencia Scripts
is a trademark of
Dodo Books Indian Ocean Ltd. and OmniScriptum S.R.L publishing group

120 High Road, East Finchley, London, N2 9ED, United Kingdom
Str. Armeneasca 28/1, office 1, Chisinau MD-2012, Republic of Moldova, Europe
Printed at: see last page
ISBN: 978-620-8-19379-9

Summary

1. introduction

Santa Catarina is Brazil's fourth largest corn producer, contributing 7% (2.3 million tons) of Brazil's production in the 2005/06 agricultural year, which was 31.8 million tons. Corn is an important crop in the agricultural sector in the state of Santa Catarina, and is grown mainly in the micro-regions of Sâo Miguel D'Oeste, Chapecó, Xanxerê, Concórdia, Joaçaba and Canoinhas (Sintese... 2005). In these regions, this cereal stands out because it is very common on farms, due to its various uses, such as the production of silage (Alves et al., 2004a), the production of feed for poultry and pig farming and the sale of grain (Testa et al., 1996).

Among the maize genotypes grown in Santa Catarina, open-pollinated varieties are among the varieties used on small farms (Cruz et al., 2006b; Goulart & Gugel, 2006; Kist et al., 2006b; Kist et al., 2006c; Canci et al., 2004). This use stems from the benefits that these varieties provide, such as adaptation to climatic conditions and cultural practices, resistance to the pests and diseases of a region (Harlan, 1975) and the use of fewer chemical inputs in crops (Jarvis et al., 2000). However, for many of the open-pollinated varieties used by farmers in Santa Catarina, there is little information on the agronomic management of these varieties, especially in relation to the optimum plant density for cultivation. Density information is essential for cultivation because the use of inadequate populations (in sub-optimal or supra-optimal conditions) directly affects the development, growth and productivity of maize due to the low phenotypic plasticity of this species (Sangoi et al., 2001a).

The general objective of this work was to study the effect of plant densities on the grain and silage yields of the open-pollinated variety MPA1. The specific objectives were to evaluate the effects of plant density on intraspecific plant competition, on phenological, morphophysiological and agronomic variables and to estimate the optimum plant density for phytomass production and grain yield.

2. Corn

2.1 The importance of growing corn

Among the species already identified in the world, cereals are the most important for satisfying our food needs (Otegui & Slafer, 2004). Among these cereals, maize stands out because it is widely used due to its excellent production potential (Fancelli & Dourado Neto, 2000) and is an excellent source of energy due to its high carbohydrate content (Steinmacher, 2005), being classified as the second most important cereal in the world, only behind wheat and followed by rice (Sintese... 2005). In addition to these characteristics, corn has several applications, ranging from the manufacture of polenta, tortillas and snacks (Steinmacher, 2005) to its use in the animal nutrition industry (EMBRAPA, 2005).

World corn production reached 692.24 million tons in the 2005/06 harvest, a reduction of approximately 3% compared to the previous world harvest (2004/05). The world's major corn producers include the United States with 42% of production, China with 18%, Brazil with 5% and Argentina and Mexico with 3% of world production (Sintese... 2005).

In Brazil, the 2005/06 crop reached 41 million tons of corn, an increase of 8.5% on the previous harvest. Among the largest Brazilian corn producing states, Paranà stands out with 24% of production, followed by Minas Gerais with 18%, São Paulo with 11%, Santa Catarina and Goiás with 8% and Rio Grande do Sul with 4% (Sintese... 2005).

In Santa Catarina, production in the 2005/06 crop year (3.343 million tons), the micro-regions of Sâo Miguel D'Oeste, Chapecó, Xanxerê, Joaçaba, Concórdia and Canoinhas were the biggest producers of this cereal. In addition, the yields achieved in the 2005/06 harvest (4,260 kg ha^{-1}) in Santa Catarina were higher than the two previous harvests of 2003/04 (4,100 kg ha-i) and 2004/05 (3,530 kg ha^{M}) (Sintese... 2005).

2.1.1 Corn on farms in Santa Catarina

Most of the maize produced in Santa Catarina comes from small farms, which is a typical characteristic of the state (Censo Agropecuàrio/SC, 1996; Sangoi et al., 2003). On these smallholdings, 58% of the cultivated areas comprise maize plantations of up to 10 hectares, where farmers consume approximately 62% of the production on their own properties, demonstrating a strong relationship between this crop and the farmers (Agricultural Census/SC, 1996).

This relationship between farmers and corn can be observed in several municipalities in Santa Catarina. Among these municipalities is the city of Anchieta, located in the far west of Santa Catarina, where 73% of rural establishments are considered to be smallholdings and where maize is the most important source of income for farmers (Vogt, 2005). In addition to the farmers' relationship with maize, this municipality also stands out for its widespread use of local and creole varieties of maize (Canci et al., 2004). As a result, it has become the "Capital of Creole Maize in Santa Catarina", according to state bill 466/99 (Canci, 2002).

These varieties cultivated in Anchieta are of great importance to farmers in the municipality and also have other important characteristics such as the genetic value of local and creole varieties, the rescue of genes for characteristics of interest, the conservation and use of genotypes by the most oppressed classes, as well as being a reliable way of producing food under adverse climate and soil conditions (Kist, 2006a).

2.2 Arrangement of plants

Maize is a crop that achieves high yields due to two fundamental characteristics associated with the species. The first is that the plants have enormous potential for grain production (Sangoi, 2000), while the second is that the species is highly responsive to the increased management applied to the crop (Sangoi et al., 2003). The response of maize to increased management was evident in North America between the 1930s and 1990s through the increase in yields, since, of the increase generated between those decades, 40% of yields were the result of agronomic management, while the rest was the result of genetic plant improvement (Tollenaar & Lee, 2002). Therefore, management techniques are fundamental for corn and have proven to be efficient tools for increasing productivity. Among the management techniques used in maize cultivation, plant arrangement stands out.

Plant arrangement is an important technique because it allows plant populations to have more adequate availability of environmental resources such as water, nutrients and solar radiation, promoting better plant development, mainly by increasing the interception and capture of solar radiation (Argenta et al., 2001). In addition, plant arrangement is fundamental for maize because this species has low phenotypic plasticity and does not have efficient mechanisms that allow plants to occupy their available niche (Sangoi et al., 2001a), due to its low tillering capacity and lack of branching (Romano, 2005).

Plant arrangement provides numerous benefits to the maize crop, but in order to identify the ideal arrangement, it is necessary to adjust the plant densities, as well as the spacing between rows and between plants (Argenta et al., 2001). Among these adjustments, plant densities stand out for being a tool that offers many possible arrangements (Sangoi, 2000). Because of these arrangement possibilities, densities were one of the main objectives of corn experiments between the 1960s and 1980s (Silva et al., 2006). In these studies, researchers sought to determine the ideal plant populations for the most diverse growing conditions and genotypes (Silva et al., 2006; Sangoi, 2000), increasing yields through the use of an optimum plant density.

2.3 Interception of solar radiation.

Solar radiation is a vital source of energy for plants (Odum, 1986; Tsumanuma, 2004) and can become a limiting factor for development in environments where radiation conditions are reduced (Odum, 1986). This limiting factor occurs because radiation has a strong relationship with the photosynthetic activity of plants, which use wavelengths between 400 and 750nm to promote the photolysis of water and the reduction of carbon dioxide (CO_2) into large carbon compounds, especially carbohydrates (Fancelli & Dourado-Netto, 2000). In maize, this reduction of carbon dioxide is carried out by a photosynthetic apparatus (via C_4) present in the plant canopy (Loomis & Connor, 1992; Maddonni et al., 2001) which transforms inorganic carbon into carbohydrates in a highly efficient manner (Silva et al., 2006; Sangoi et al., 2002a). Therefore, the use of adequate plant densities in maize crops is fundamental for obtaining good yields, as these densities promote increased collectivity between plants (Sangoi et al., 2001a, Loomis & Connor, 1992) and optimize the interception and capture of incident solar radiation (Marchâo et al., 2006, Argenta et al., 2001; Loomis & Connor, 1992; Clements, 1963).

In addition, the use of optimum densities in yarrow promotes increased interception of solar radiation (Andrade et al., 1993), increasing the amount of energy available to form carbohydrates (Sangoi, 2000), allowing greater carbon accumulation (Marchâo et al., 2006) and providing plants with greater increases in phytomass under conditions of high radiation (Didonet et al., 2002, Silva et al., 2006).

In crops where plant densities are inadequate (sub-optimal or supra-optimal), there are negative effects on the ability to intercept radiation and the production of phytomass (Sangoi, 2000). Among the possible negative effects generated by the use of sub-optimal densities is the decrease in interception (Andrade et al., 1993) and absorption of photosynthetically active radiation (RFA) (Tollenaar & Bruulsema, 1988) caused by the reduction in the leaf area index of these populations. The reduction in absorbed solar radiation from 34.9 mol photon m^{-2} d^{-1} at a density of 0,000 plants ha^{-1} to 31.5 mol photon m d^{-2-1} at a density of 39,000 plants ha^{-1} was observed by Tollenaar & Bruulsema,

(1988).

On the other hand, in crops where plant densities are above the optimum densities (supra-optimum), there is an increase in solar radiation interception as a result of the increase in leaf area index. However, this increase does not necessarily result in an increase in productivity, because at high densities the shading between plants increases, causing a reduction in the individual capacity of these plants to intercept radiation (Didonet et al., 2002, Andrade et al., 1993) and consequently convert it into phytomass.

2.4 Interspecies competition

In plant cultivation, there are interactions that affect plant growth and development. One of these interactions refers to the proximity of individuals in a population competing with each other for environmental resources such as water, light and nutrients. This type of competition, known as intraspecific competition, is not normally identified during the initial growth phase, because the plants are small at this time and do not interfere with each other's development. However, as soon as one plant interferes with another, intraspecific competition begins (Loomis & Connor 1992), which can occur depending on the plant densities used and the arrangement of the plants in the field (Silva et al., 2006). Increased competition causes an increase in the extreme red/red ratio of a population, so that the growth of stems and leaves can be altered by this ratio (Taiz & Zeiger, 2004). The red/extreme red ratio in suitable conditions for plants provides more balanced endogenous (hormonal) conditions, which allow for better allometric growth of plant organs, enabling the generation of individuals with a better capacity to intercept incident radiation (Silva et al., 2004).

Interspecific plant competition can lead to changes in the growth and development of individual plants, generating alterations in morphology, physiology, phytomass formation and accumulation (Durães et al., 1995; Loomis & Connor 1992) and grain yield (Silva et al., 2006). In addition to these alterations, there may also be changes in the individual plants' ability to seek out resources in the environment, so that different levels of competition arise within a population (Maddonni & Otegui, 2006) as a result of increased plant densities (Silva et al., 2006; Tollenaar et al., 2000; Odum, 1986). These levels of competition in populations are generated by the different individual abilities of plants to obtain their resources, which can be classified as dominant or dominated individuals (Maddonni & Otegui, 2006).

The changes caused by density have been minimized in recent decades through changes in plant architecture (hybrids), mainly through a reduction in size. This has reduced the effects of intraspecific competition (Tollenaar et al., 1997) and increased the tolerance of individuals to radiation stress (Tollenaar & Wu 1999).

2.5 Corn phenology

Maize is a crop that has five basic stages of development in its biological cycle, consisting of germination/emergence, vegetative growth, male/female flowering, fruiting and physiological maturation (Fancelli & Dourado Neto, 2000). However, in order to facilitate management and studies of maize, the biological cycle of this crop has been divided into two stages called vegetative development and reproductive development (Ritchie et al., 2003).

The biological cycle of maize can vary in terms of the number of days between emergence and physiological maturity of the varieties, allowing them to be classified from super early varieties (90 days) to highly late varieties (300 days) (Fancelli & Dourado Neto, 2000). Another variation in the maize cycle is related to the increased intervals between phenological stages as a result of the management applied to the crop. Among these management conditions is plant density which, depending on the populations used, leads plants to alter the intervals between the crop's phenological stages (Tokatlidis & Koutroubas 2004; Sangoi et al., 2002[a] , Sangoi et al., 2001a, Tollenaar et al., 2000, Otegui, 1997, Durâes et al., 1995). These changes include an increase in the number of days between emergence and male flowering in maize (Sangoi et al., 2002c). Densities containing 25,000 and 100,000 plants ha^{-1} increased the average number of days between emergence and male flowering from 68 days to 72 days respectively among the different hybrids evaluated (Sangoi et al., 2002c). Similar effects on male flowering were identified by Tokatlidis & Koutroubas (2003).

Another variation caused by plant densities is the increase in the interval of days between male and female flowering. This interval between blooms can be lengthened by up to 1.02 days for every 10,000 plants added to^{-1} (Sangoi et al., 2002c). Tokatlidis & Koutroubas (2004) described that by increasing densities from 50,000 to 200,000 plants ha^{-1} , the interval between flowering was increased by up to 9 days.

In addition to delaying male and female flowering in maize, plant densities also cause changes in the grain filling period, which deserves attention because it is directly related to grain yield (Durâes et al., 2005). These changes are the result of increased plant densities reducing the grain filling period. This reduction was verified for a late hybrid grown at plant densities of 50,000, 75,000 and 100,000 plants ha^{-1} , where increasing densities caused a reduction of up to 12 days in the grain filling period (Sangoi et al., 2002 d). These effects of densities on the reduction of the grain filling period were also identified by Maddonni et al. (2001).

2.6 Plant architecture

In order for the maize crop to be able to grow at high plant densities and optimize the interception and capture of incident solar radiation without altering the photosynthetic rates of the plants, some changes have been made to the architecture of older maize varieties because the materials are extremely sensitive to high plant densities. These changes in varieties began in the second half of the 20th century and led to considerable progress in many hybrids (Sangoi et al., 2002a), making these (modern) materials more tolerant of high plant densities (Fellner et al., 2003). Among the changes brought about in plant architecture by genetic improvement is a reduction in canopy size and an increase in leaf angle (Tollenaar et al., 2000, Fellner et al., 2003). These two changes have allowed plant canopies to become smaller and with more erect leaves, occupying less space per individual and thus allowing plant densities to increase (Duvick & Cassman, 1999).

The development of new genotypes with architectures tolerant of high densities has benefited the maize crop by reducing the number of sterile plants, increasing the synchrony between male and female flowering, increasing the productive potential of hybrids (Sangoi et al., 2002a), improved the ability to use (Kamara et al., 2004) and intercept incident radiation (Marchâo et al., 2006) and increased energy storage capacity with minimal wear and tear on vegetation (Sangoi, 2000).

However, the morphological, phenological (Kamara et al., 2004) and physiological (Sangoi, 2000) changes in the maize crop have not altered the photosynthetic rates of the plants. These changes have allowed the new hybrids to obtain photosynthetic rates that are less susceptible to the stresses of high plant densities when compared to the photosynthetic rates of the old varieties (Tollenaar et al., 2000).

Old varieties and open-pollinated varieties, also known as local and creole varieties, have contrasting characteristics to those of hybrids (Balbinot et al., 2005). They are more susceptible to the effects of densities because they have higher plant height and ear insertion than the average height identified among improved varieties (Balbinot et al., 2003), 2003). They also have a greater leaf angle and a larger total leaf area per plant, which give the individuals more open leaves, generating a larger canopy than the improved varieties.

2.7 Changes in the morphology of maize plants

When maize plants are grown at high plant densities, they undergo some variations in their morphology (Sangoi et al., 2002b; Durâes et al., 1995).

Among the main morphological variations caused by the effect of high densities are variations in stalk diameter, plant height, ear insertion height, number of leaves, number of ears per plant, leaf angle and leaf area.

2.7.1 Thatch

The corn stalk becomes visible shortly after emergence when the plants have 3 expanded leaves, and the development of this organ accelerates with the appearance of the 6th leaf. From the 30th day after emergence (DAE) until close to flowering, the stalk undergoes a period of great growth and elongation with a significant accumulation of phytomass, followed by a small reduction until the end of the crop (Ritchie et al., 2003) which can reach 14 grams plant^{-1} (Sangoi et al., 2002a).

The reduction in stalk diameter after flowering is caused by the remobilization of this organ's reserve substances for grain filling (Sangoi et al., 2002a). For this reason, the stalks are considered to be very important modulating structures for grain yield, especially when the leaves suffer some kind of damage during the reproductive phase of the crop (Sangoi et al., 2001a). This organ can be defined as a structure that balances source limitation by promoting the remobilization of stored reserve carbohydrates until the beginning of grain filling (Uhart & Andrade, 1995).

In addition to the reduction in stalks caused by the remobilization of accumulated carbohydrates, the stalk is among the maize structures most susceptible to the effects of intraspecific competition generated by the effects of plant density. The diameter of the stalks at densities of between 25,000 and 100,000 plants ha^{-1} was reduced by 33% (26.9 mm and 18.0 mm) as a result of the increase in density (Sangoi et al., 2001a).

The natural reduction in stalk diameter that occurs after flowering, together with the reduction caused by the effect of density, are responsible for the fragility generated in this organ (Costa et al., 2005, Sangoi et al., 2002b, Argenta et al., 2001, Sangoi et al., 2001a) leading to an increase in the percentage of lodged and broken plants. Both lodging and breakage are variables susceptible to the effect of plant density (Marchâo et al., 2006). At densities of 25,000, 50,000, 75,000 and 100,000 plants ha- 1, lodging and plant breakage reached percentages of 0.5%, 3.6%, 5.9% and 12.6%, respectively (Sangoi et al., 2001a).

2.7.2 Spike insertion height

Ear insertion height refers to the distance between the neck of the plant and the base where the main ear is inserted. Among existing varieties, ear insertion height varies according to the genotype. Ear insertion height can vary between 120 and 200 cm for open-pollinated varieties (ALVES et al., 2004b) and between 92 and 141 cm (Kamara et al., 2004) and 104 and 221 cm (Balbinot et al., 2003) for hybrids.

In the same way that varieties show differences in ear insertion height, a single maize variety grown at different plant densities can also show significant differences in ear insertion height (Sangoi et al., 2002c). This information was confirmed by Sangoi et al. (2002b) who used densities of 25,000, 50,000, 75,000 and 100,000 plants ha^{-1} and obtained ear insertion heights of 123, 132, 137 and 137 cm respectively. Thus, ear insertion height increases with increasing plant densities (Sangoi et al., 2002c). A significant linear increase was also identified with an increase in plant density from 40,000 to 100,000 plants ha- 1 (Marchâo et al., 2006).

On the other hand, Borghi et al. (2004) and Silva et al. (1999), using densities varying between 50,000 and 100,000 plants ha- ı, did not identify any changes in the height at which the ears were inserted, showing homogeneous results between the densities.

The lower height of the ears is advantageous because it allows the plants to remain upright until harvest due to the shorter distance between the insertion of the ear and the ground. In addition, they contribute to better plant balance by minimizing stalk breakage, especially at high densities where the diameter of this structure is smaller (Sangoi et al., 2002b). On the other hand, insertion height can damage plants when the ear insertion height is high. This is because high ear insertion causes problems with lodging and plant breakage as a result of the increased intensity of the pendulum formed between the ear and the ground, caused by the greater displacement of the individuals' center of gravity (Sangoi et al., 2002b). In addition, the conditions of high plant densities lead to a reduction in the diameter of the stalk which, added to the displacement of the plant's center of gravity, favors stalk breakage (Sangoi et al., 2001a).

2.7.3 Plant height

Some studies carried out with maize showed no differences in plant height between individuals grown at different plant densities (Costa et al., 2005, Borgui et al., 2004, Silva et al., 1999, Durâes et al., 1995). However, other studies have reported significant changes in plant height as a result of increasing plant density (Marchâo et al., 2006, Sangoi et al., 2002c; Maddonni et al., 2001; Sangoi & Salvador, 1996; Modarres et al., 1998). Sangoi & Salvador (1996) described the increase in densities from 25,000 to 75,000 plants ha^{-1} where plant height went from 225 cm to 236 cm, as did Sangoi et al. (2002b) who identified an increase of 276, 282, 286 and 280 cm in populations of 25,000, 50,000, 75,000 and 100,000 plants ha- 1, respectively.

The increase in plant height under conditions of greater densification is due to the plants showing a tendency to grow shorter as the incident radiation on the individuals decreases, indicating greater intraspecific competition for light (Sangoi et al., 2002c; Modarres et al., 1998). The lower incidence of radiation on the individuals causes an increase in endogenous auxin levels (greater shading between plants and consequently lower auxin photoxidation), promoting stem elongation and an increase in plant height (Taiz & Zeiger, 2004).

2.7.4 Total number of leaves

Plant densities of 50,000 and 90,000 plants ha^{-1} (Otegui, 1997) and 25,000 and 100,000 plants ha^{-1} (Sangoi et al., 2002 b) showed no change in the total number of leaves present on the plants (21 and 21.3 leaves, respectively). Contradictory to this information, a reduction in the number of leaves above the main ear was identified as a result of increasing plant densities from 65,000 to 90,000 plants ha^{-1}, where there was a reduction from 9 to 7 leaves respectively (Modarres et al., 1998).

2.7.5 Leaf angle

The leaf angle is the angle between the main vein of the leaf and the stalk of the plant. This angle can vary between maize varieties due to the intrinsic characteristics of each genotype (Maddonni & Otegui, 1996). However, these variations in leaf angle can also be the result of using different plant densities (Silva et al., 2006 and Drouet & Moulia, 1997). Maddonni et al. (2001) found a 10% increase in the angle of leaf insertion (50.5° to 55°) in the upper stratum of the plants (above the ear) as a result of the increase in plant density.

2.7.6 Leaf area

The increase in plant density promotes an increase in the leaf area index of the corn crop (Mendes et al., 2005, Lindquist et al., 2005, Almeida et al., 2003, Gardiol et al., 2003, Silva et al., 1999). In studies where densities were increased from 39,000 to 90,000 plants ha^{-1} (Cox, 1996); 39,000 to 100,000 plants ha^{-1} (Tollenaar & Bruulsema 1988) and from 30,000 to 120,000 plants ha^{-1} the leaf area index increased by 72% (2.50 - 4.30), 90% (2.40 - 4.56) and 309% (2.19 - 6.78), respectively.

On the other hand, when the leaf area per plant is evaluated, it can be seen that an increase in density leads to a reduction in the total leaf area per plant as a result of increased intraspecific competition. At plant densities of 25,000, 50,000, 75,000 and 100,000 plants ha^{-1}, total leaf area per plant was 9,879;

8,6407; 7,355 and 6,482 cm^2 respectively among the hybrids evaluated (Sangoi et al., 2002b), demonstrating a reduction in total leaf area per plant as density increased.

The individual areas of the leaves are also subject to reduction as a result of the increase in plant density. At densities of 90,000 and 120,000 plants ha- 1, a 21% reduction in leaf width and a 26% reduction in leaf length were observed compared to the leaves of plants grown at a density of 30,000 plants ha^{-1} (Maddonni et al., 2001). This reduction in area caused by increased intraspecific competition was also identified by Sangoi et al. (2002c) and Modarres et al. (1998). Another variable affected by density is senescent area per plant, where an increase in density leads to an increase in senescent area per plant, especially after flowering (Borràs et al., 2003).

2.8 Phytomass production

The production of corn phytomass per area and per plant can be altered when crops are subjected to different densities. These changes occur because the increase in the number of individuals per area (Durâes et al., 1995; Borghi et al., 2004; Weddicombe & Thelen, 2002; Tollenaar & Bruulsema 1988) increases the production of phytomass per unit area (Costa et al., 2005; Cox, 1996; Tollenaar & Bruulsema, 1988) and decreases the amount of phytomass produced per plant.

Increasing densities from 33,000 to 77,000 plants ha^{-1} (Durâes et al., 1995), 39,000 to 100,000 plants ha- ı (Tollenaar & Bruulsema 1988) and 66,000 to 121,000 plants ha- 1 (Costa et al., 2005) promoted an increase in phytomass from 11,110 to 18,440 kg ha- 1, 15,000 kg ha- 1 to 17,400 kg ha- 1 and 15,000 to 20,000 kg ha- 1, respectively. Cox (1996), evaluating younger plants (12 leaves), also found a significant upward trend in the phytomass produced (6,750 kg ha- 1 to 9,350 kg ha- 1) caused by increasing densities from 30,000 to 90,000 plants ha 1-

The increase in phytomass production per unit area is not linearly proportional to the increase in plant density (Costa et al., 2005). Despite this, higher densities produce more phytomass per area than lower densities (Cox, 1996; Durães et al., 1995). This increase is characterized by occurring up to the optimum cultivation density established for a variety, and this optimum density varies depending on the genotypes used, as well as biotic and abiotic factors. For populations grown in conditions higher than the established optimum density, considerable reductions are observed in the phytomass produced per area, which generates a quadratic response of this variable (Durães et al., 1995; Cox, 1996; Tollenaar et al., 1992; Tokatlidis, 2001). The decrease in phytomass production per area results from the fact that individuals grown in denser populations show a large decrease in the amount of individual phytomass, and this decrease is more pronounced than the increase promoted by the increase in plant density. Individuals grown at densities of 25,000 and 83,000 plants ha^{-1} had an average reduction of 40% in the phytomass produced per plant (311123 g) (Tokatlidis, 2001).

2.9 Grain yield

Grain yield is significantly altered by the effect of plant density. These changes occur because increasing densities increase grain yield (Maddonni & Otegui, 2006; Sangoi et al., 2006a; Sangoi et al., 2002d; Sangoi et al., 2002c; Sangoi et al., 2002b; Sangoi et al., 2001 a; Tokatlidis 2001) in such a way that this elevation shows linear (Sangoi et al., 2001a) or quadratic (Flesch & Vieira, 2004; Silva et al., 1999) behavior. These linear or quadratic behaviors probably result from the range of plant densities used (Table 1).

Table 1. Plant density, grain yield and source of some studies on plant density.

Plant densities (plants ha^{-1}) Minimum - maximum	Grain yield (Kg ha^{-1}) Minimum - maximum	Source
35.000 - 80.000	7.500 - 10.500	Almeida et al. (2000)
55.000 - 75.000	4.584 - 5.472	Borghi et al. (2004)

45.000 - 90.000	7.200 - 8.200	Cox (1996)
33.000 - 77.000	4.760 - 7.810	Durâes et al. (1995)
30.000 - 73.000	6.000 - 7.448	Flesch & Vieira (2004)
50.000 - 70.000	8.300 - 9.100	Silva et al. (1999)
25.000 - 75.000	6.846 - 14.461	Sangoi & Salvador (1996)
39.000 - 100.000	6.540 - 7.800	Tollenaar & Bruulsema (1988)

Among the main yield components affected by plant density are the weight of 1000 grains and the number of grains per area. The weight of 1000 grains interferes with productivity, because as plant densities increase, the efficiency with which plants convert these photoassimilates into grains is reduced (Sangoi et al., 2002a), leading to a decrease in the weight of 1000 grains (Maddonni & Otegui, 2006; Hashemi et al., 2005; Sangoi & Salvador, 1996; Borghi et al., 2004; Flesch & Vieira, 2004; Echarte et al., 2000). This reduction in conversion efficiency became evident with the increase in plant densities from 45,000 plants ha^{-1} to 90,000 plants ha^{-1} where there was a reduction from 336 g to 306 g in the weight of 1000 grains (Cox, 1996). In contrast, Sangoi et al. (2006b) found no differences in the weight of 1000 grains for an open-pollinated variety grown at densities ranging from 40,000 to 60,000 plants m .$^{-2}$

In relation to the number of grains per unit area, there is an increase in this variable in response to densification caused by an increase in the interception of photosynthetically active radiation per plant (Kiniry & Knievel, 1995). Increasing densities from 46,000 plants ha^{-1} to 70,000, 84,000 and 93,000 plants ha^{-1} increased the number of grains per area from 2,440 to 3,320, 3,750 and 3,880 grains, respectively (Andrade et al., 1993). Thus, the increase in the number of grains per area affects productivity (Durâes et al., 2005). However, this behavior occurs until the maximum number of grains is reached, because from then on, high plant densities (above the optimum densities) lead to a reduction in the number of ears per plant (Cox, 1996; Tollenaar et al., 1992), the number of kernels per plant (Maddonni & Otegui, 2006; Otegui, 1997;

Echarte et al., 2000; Cox, 1996; Sangoi & Salvador, 1996; Tollenaar et al., 1992 and Jacobs & Pearson, 1991), the number of kernels per ear (Silva et al., 1999) and consequently the number of kernels per area.

The reduction in the number of grains per area and ears per plant is evident in two populations containing 45,000 and 90,000 plants ha^{-1}. At a plant density of 45,000 plants ha^{-1}, 523 grains were found m^{-2} and 1.02 ears per $plant^{-1}$, while at a density of 90,000 plants ha^{-1} this number fell to 383 grains m^{-2} and 0.96 ears per $plant^{-1}$ (Cox 1996). Similarly, a reduction from 428 to 379 and 315 grains per ear was found as densities increased from 50,000 to 70,000 and 90,000 plants ha^{-1}, respectively (Silva et al., 1999). Flesch & Vieira (2004) also identified a reduction in the number of kernels per ear (547, 487, 394 and 344) and the number of ears per plant (1.05, 1.01, 0.97 and 0.95) with increasing densities (30,000, 50,000, 70,000 and 90,000 plants ha^{-1}) for a normal cycle hybrid.

2.10 harvest index

The harvest index of a variety can be altered by plant densities because densities cause changes (increases or decreases) in the yield components. Changes in the components (increase or decrease) alter the ratio between total phytomass and grain phytomass, making the harvest index susceptible to variations depending on the plant density used (Maddonni & Otegui, 2006; Hashemi et al., 2005).

In addition to changes in yield components, plant densities can also cause changes in the amount of total phytomass produced. This total phytomass can follow the increase in plant density, so that a simultaneous increase in grain yield and total phytomass does not cause changes in the maize harvest index. This behavior of total phytomass and kernel yield was observed in the plant densities of 45,000, 67,000 and 90,000 plants ha^{-1} used by Cox, (1996), where an increase in phytomass and kernels was observed, however, no changes were identified in the harvest index of the hybrid used, which remained at 0.49

(Cox, 1996). On the other hand, Borghi et al. (2004) using plant densities of 55,000, 65,000 and 75,000 plants ha^{-1} obtained harvest indices of 0.56, 0.56 and 0.62 respectively for the hybrid used, demonstrating a relationship between plant density and harvest index. Sangoi et al., (2001a) using plant densities of 25,000, 50,000, 75,000 and 100,000 plants ha^{-1} and plant material with contrasting cycles, obtained harvest indices of 0.55; 0.47 and 0.35 for super early, early and late hybrids respectively. These harvest indices showed the influence of the plant cycle on the harvest index and that the index was not related to plant density.

3. Material and methods

3.1 Location

The experiment was carried out at Epagri's Campos Novos Experimental Station, located in the midwest of Santa Catarina at an altitude of 934 meters above sea level, latitude -27° 24' 0" and longitude -51° 13' 30" (CIASC, 2005).

3.2 Weather

The predominant climate in Campos Novos is temperate with low winter temperatures. Classified according to Koeppen as Cfb temperate (Pandolfo et al., 2002).

3.3 Soil

The soil at the site of the experiment is classified as Nitossolo, with good drainage characteristics and a depth normally ranging from 1.5 to 2.5 meters, with the "A" horizon being between 25 and 55 centimeters thick. The station's soils are located on the lower third of a slope, with a 14% gradient, and show moderate erosion (Laus Neto et al., 1999).

3.4 Plant material

The plant material used was MPA1 corn compost produced by farmer Névio Alceu Folgiarini, from the Linha Sâo Roque community in the municipality of Anchieta/SC, with the technical support of Adriano Canci from the Association of Small Producers of Organic Creole Corn and derivatives of Anchieta and the Union of Family Agriculture Workers of this municipality (Kist, 2006a).

The development of this compound took place in the 1999/2000 agricultural year, when 25 different maize populations were initially brought together, of which 18 were commercial synthetics, four were open-pollinated varieties of the pixurum group and three were local or creole varieties (Cateto, Mato Grosso, Palha Roxa and Amarelâo) grown in Anchieta (Kist, 2006a). On farmer Névio Alceu Flogiarini's property, an isolated area was set aside for recombination between the 25 populations. In this case, under a plant density of 40,000 plants ha^{-1} , a 40-meter row made up of an equitable mixture of seeds from all 25 populations was interspersed with two individual rows, each represented by one of the 25 populations. The rows represented by the mixture of the 25 populations formed the male rows, while the single rows made up individually of one population were peeled and formed the female rows (Kist, 2006a).

The seed-providing plants were selected every 10 linear meters by choosing five plants based on their expression of characters relating to plant height, prolificacy, stalk diameter and disease resistance. From the total area, 500 plants were selected from which 500 ears were obtained. In the post-harvest phase, the best ears were selected based on ear size, degree of stuffing, type and color of kernels (Kist, 2006a).

Another five unknown populations were included in the 2000/2001 crop year. The composite resulting from the recombination of the 30 populations produced 300 ears which were used to produce the seeds of the composite population MPA1. In the following three harvests, three cycles of stratified Massai selection were carried out according to Soares et al. (1998), from which the seeds of the MPA1 composite population were produced (Kist, 2006a).

3.5 Experimental design

The experimental design used was completely randomized blocks with four replications and five treatments consisting of (average) densities of 18,000, 34,000, 47,000, 56,000, 75,000 plants ha $.^{-1}$

The experiment consisted of 20 plots, each 8 meters wide, 7 meters long and spaced 1 meter apart, giving a total plot size of 56 m^2 and a total experimental area of 1,120 m2. The experimental area was demarcated on September 30, 2005.

3.6 Conducting the experiment

3.6.1 Agro-ecological characterization of the area

The area at the experimental station has been managed using the agroecological farming system since 2000. During the first two years, the area went through a period of conversion, with the first direct planting (soybeans) only taking place in the second half of 2002. The following year (2003), the area remained fallow, and in 2004, there was a new direct planting of soybeans. When these crops were planted, some safety measures were taken, such as the presence of margins of approximately 10 meters at the ends of the area to avoid proximity to conventional plantations.

During the period in which the area remained fallow, only spontaneous plants grew, predominantly species such as papuà (*Brachiaria plantagina*), ryegrass (*Lolium multiforum*), prickly pear (*Bidens pilosa*) and yarrow (*Euphorbia sp*). The fertilizer commonly used on the area was green manure and organic fertilizers such as poultry litter (5,000 kg ha).$^{-1}$

3.6.2 Soil preparation and fertilization

Soil preparation consisted of plowing and harrowing. After preparation, soil samples were taken from the area of the experiment and sent for analysis to the Epagri laboratory in Chapecó.

The area was fertilized by adding 14,000 kg ha^{-1} (dry basis) of poultry manure, distributed at the same time on the plots and then incorporated into the soil by harrowing. At 49 days after emergence (DAE), the crop was fertilized by spreading 8,500 kg ha^{-1} (dry basis) of poultry manure at the base of the plants.

The amount of organic fertilizer used was determined taking into account the results of the soil analysis (Table 2) and the characteristics of the fertilizer used (Table 3).

Table 2. Results of the soil analysis of the experimental area.

Block	%Clay m/v	pH -Water	SMP index	P mg/dm³	K mg/dm3	% MO m/v	Al cmolc/dm3
1	65	5,6	5,9	18,0	400	5,2	0
2	69	5,9	6,1	7,9	339	4,6	0
3	66	5,7	6,0	9,7	351	5,2	0
4	70	5,4	5,8	7,0	340	4,0	0

Block	Ca	Mg	H + Al	CTC	% Saturation	
	cmolc/dm³	cmolc/dm3 c	molc/dm3ⁱ	cmolc/dm3	Bases	Al
1	9,8	4,1	4,89	19,81	75,32	0
2	10,3	4,4	3,89	19,46	80,01	0
3	10,4	3,8	4,36	19,46	77,59	0
4	8,2	3,5	5,49	18,06	69,60	0

Table 3. Analytical results of organic fertilizer (poultry manure).

pH	M.S	N	P	KCa	Mg
	%			%	
9	74,44	3,06	1,25	2,043 ,9	1,36

3.6.3 Crop management

Sowing was done manually in a row on November 3, 2005. The amount sown at plant densities of 47,000, 56,000 and 75,000 plants ha^{-1} was twice the number of plants calculated for the stands, while for the populations of 18,000 and 34,000 plants ha^{-1} this amount was three times greater. The plants were thinned out at 10 DAE, adjusting the ideal number of plants to their respective treatments.

In order to control spontaneous plants that could compromise the development of the corn in the initial phase, two weeding sessions were carried out throughout the growing season, the first at 16 DAE (30/11/2005) and the second at 53 DAE (06/01/2006). From this date onwards, no more activities were carried out to control spontaneous plants. No pest or disease control was carried out, given that the experiment was based on an agro-ecological system of cultivation.

After the experiment was sown, there was a drought in the Campos Novos region and the accumulated rainfall during this period was insufficient for the corn plants to develop well. Four irrigations were therefore necessary in the experimental area, each equivalent to 32mm at 28, 32, 36 and 50 DAE.

3.7 Phenology

Four phenological stages of the MPA1 variety were assessed: emergence (appearance of the coleoptile), male flowering (FM), female flowering (FF) and physiological maturation of the kernels (MF) according to the phenological stages described by Ritchie et al. (2003). The date of emergence was determined when the number of emerged seedlings in the 4th row reached 50% of the plants in the row.

Male flowering was determined when the last branch of the pendant became visible and pollination of the variety began (Ritchie et al., 2003). Female flowering was determined when the stigma of the ears became visible. Thus, the DAE of the blooms were established through evaluations made on rows 6 and 7, which represent a useful area of 12 m^2 . These evaluations were carried out every three days, identifying the number of plants with the characteristic in the area, in order to determine when 50% of the blooms had occurred.

Physiological ripeness (stage R6) occurred when the kernels showed the black layer of abscission (Ritchie et al., 2003). To determine this phenological stage, the number of ears present in row number 4 of each plot was assessed and the ears were evaluated until at least 50% of the individuals showed this characteristic.

The estimate of 50% of male and female flowering and physiological maturation was based on a linear regression equation between the percentages of flowering, physiological maturation and the number of days after emergence.

3.8 Morphology

To determine morphological variables such as plant height, stem diameter and number of leaves (expanded, visible, green and senescent), 4 plants (rows 6 and 7) were chosen and marked in each plot. These individuals were evaluated from the first week until the end of the MPA1 compost cycle, and these evaluations were carried out weekly until male flowering. After this phenological stage, evaluations were carried out every 14 days until the corn was harvested.

For variables such as ear height and leaf angle, the same plants were used for the evaluations, but these variables were evaluated once during the cycle. The height of the main ear had its data obtained at the physiological maturity of the crop and the leaf angle was evaluated at male flowering.

3.8.1 Plant height

Plant height was measured using the distance between the neck of the plant and the highest point of the leaves. To obtain this variable, a 400 cm ruler was used.

3.8.2 Spike insertion height

The height of the insertion of the ear was measured using the distance between the neck of the plant and the insertion of the main ear. This reading was obtained using a graduated ruler.

3.8.3 Stalk diameter

Stalk diameter was measured at an average height of 5 cm above ground level. A digital caliper was used to determine this variable.

3.8.4 Number of leaves

The number of expanded, visible, green and senescent leaves was obtained through weekly leaf assessments on previously marked plants. These plants had 5ª , 10ª and 15ª leaves identified for subsequent counting.

3.8.5 Leaf angle

This variable was measured using the angle between the stalk of the plants and the central vein of the first leaf above the main ear, according to the national system for protecting cultivars (Ministério... 1997). The same evaluation procedure was used to determine the angle of the 12th leaf. A protractor was used to determine this variable.

3.9 Leaf area

The variables individual leaf area, total leaf area per plant, senescent area per plant, green area per plant and leaf area index were calculated by taking readings of the length and width of expanded leaves on all the leaves present on the four plants previously marked in rows 6 and 7 of each plot.

3.9.1 Individual leaf area

To determine the individual leaf area, four plants per plot were marked and the length (C) and width (L) of the expanded leaves present on the plants were measured. The individual leaf area of each leaf was calculated using the formula suggested by Tollenaar (1992) where:

AFI = L x W x W.

where: W is the leaf correction factor and a value of 0.75 was used (Tollenaar, 1992).

3.9.2 Total leaf area per plant

The total leaf area per plant was calculated by adding up the individual leaf areas of the plants evaluated.

3.9.3 Green area per plant

The green area per plant was calculated by adding up the leaf areas of the expanded and visible leaves, minus the senescent leaf area. To determine the area of the visible leaves, it was established that 2 visible leaves are equivalent to the area of one expanded leaf.

3.9.4 Relative senescent area per plant

The relative senescent area per plant was obtained from the ratio between the senescent area per plant and the total area of each plant.

3.9.5 leaf area index

The leaf area index (LAI) was calculated according to the formula described by Almeida et al. (2003) where:

IAF = AVPP x NP

where: AVPP refers to the green area per plant and NP to the number of plants m $.^{-2}$

3.10 Agronomic characteristics

To determine phytomass, plants were collected from 2 linear meters of each plot at 1, 11, 19, 49, 84, 112 and 154 DAE. When the material was harvested at 197 DAE, the last collection was made in a useful area of 12 m2 in rows 6 and 7. The number of ears per plant, number of ears m^{-} 2, number of kernels m^{-} 2, weight of 1000 kernels, kernel yield, harvest index and lodging and plant breakage were determined in this same 12 m2 area.

3.10.1 Phytomass

Total phytomass, stem phytomass, proportion of stems per plant, green leaf phytomass, proportion of green leaves per plant, senescent leaf phytomass, pendant phytomass and cob and grain phytomass were determined using sub-samples of 4 plants taken from the total number of individuals collected (totaling 16 plants per treatment). These plants were dissected into stems, green leaves, senescent leaves, pendants, spikes and kernels (if present) and placed in a drying oven at 65°C until their weight stabilized.

3.10.2 Number of ears planted^{-1}

The number of spikes per plant was calculated by taking the total number of spikes collected in a useful area of 12 m^2 (rows 6 and 7), divided by the number of individuals present in that area.

3.10.3 Number of ears m^{-2}

The number of ears $_{m-2}$ was calculated using the total number of ears collected in the 12 m2 area (rows 6 and 7) divided by the collection area.

3.10.4 Number of grains m 2^{-}

The number of kernels in^{-} 2 was calculated using the weight of kernels obtained in 1 m2, divided by the dry weight of one kernel.

3.10.5 Number of grains in the ear^{-} ı

The number of kernels per ear was obtained by dividing the number of kernels per area by the number of ears in the same area.

3.10.6 Weight 1000 grams

The weight of 1000 grains was obtained by averaging the weight of 8 sub-samples of 100 grains and multiplying by 10.

3.10.7 Grain yield

Grain yield (kg ha^{-1}) was obtained from the total weight of grains harvested from the 12 m $plot^2$ and corrected to 13% moisture.

3.10.8 harvest index

The harvest index was calculated by dividing the grain phytomass by the total phytomass.

3.10.9 Plant lodging and breakage

Evaluations of lodging and plant breakage were carried out on a 12 m2 area, where the total number of lodged and broken plants was determined at 155 DAE (18/04/2006) and 196 DAE (29/05/2006 harvest). The criterion used to assess lodged plants was those with an angle of inclination of the stalk greater than 45° and broken plants were those in which the stalk was twisted, cracked or broken below the point of insertion of the ear (Sangoi et al., 2002 b).

3.10.10 Spontaneous plant production

The phytomass of spontaneous plants between the corn rows was quantified by collecting them from a 2 m2 area at 139 DAE (91 days after the last weeding). During this collection, all the spontaneous plants in the area were cut off at ground level and the material stored in plastic bags. The samples were then sent to the Epagri laboratory, where they were placed in a drying oven at an average temperature of 60°C for the period necessary for the weight of the phytomass to stabilize. After drying the plant material, the dry weight of the collected samples was determined using a digital precision scale.

3.11 Data analysis

The data obtained from the five plant densities studied were subjected to analysis of variance with the Duncan test ($p < 5\%$) separating the means. The significant results were then submitted to regression and correlation analysis at a 5% significance level. The regression equations and determination coefficients are shown in the graphs when they are significant at 5%.

4. Results

4.1 Phenology

The seedlings emerged 11 days after sowing and there were no differences between the plant densities used. However, there was a significant increase in the number of days from emergence to male and female flowering as the density increased (Figure 1 and Appendix 1). On average, the populations with 18,000 plants ha^{-1} had male flowering occurring at 79 DAE, while in the populations with 75,000 plants ha^{-1} male flowering occurred at 83 DAE. Female flowering in the populations with 18,000 plants ha^{-1} occurred at 82 DAE, while in the populations with 75,000 plants ha^{-1} it occurred at 91 DAE.

The increase in plant densities also caused a significant increase in the interval of days between male and female flowering (Figure 1 and Appendix 1). At densities of 18,000 plants ha^{-1} the interval was 3 days between male and female flowering, while at densities of 75,000 plants ha^{-1} this interval was 9 days.

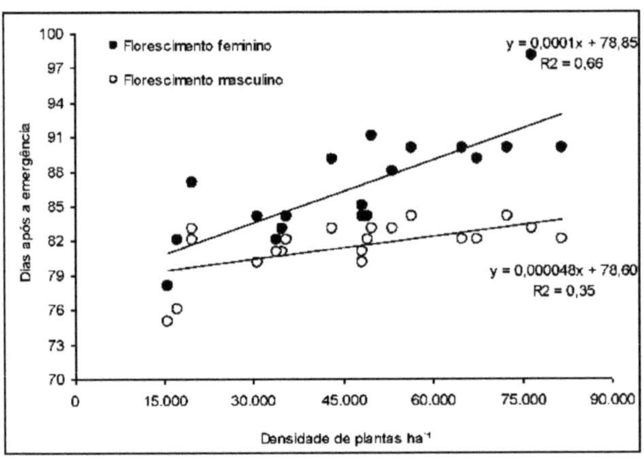

Figura 1. Number of days between emergence and male and female flowering as a function of **plant density** for maize variety MPA1.

There was a significant decrease in the number of days between female flowering and physiological maturity as plant densities increased (Figure 2 and

Appendix 2).

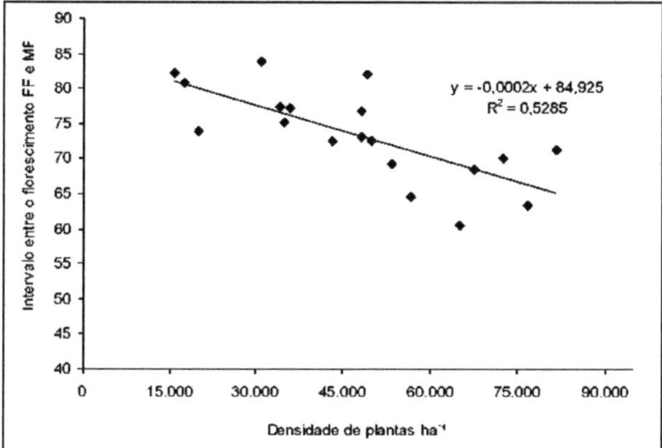

Figura 2. Number of days between female flowering and physiological maturity as a function of plant density for the MPA1 maize variety.

4.2 Morphophysiological characteristics

4.2.1 Plant height and ear insertion

During the crop cycle, plant height was characterized by two distinct phases of development (Figure 3). The first phase was characterized by almost linear growth between emergence and male flowering (approximately 81 DAE), while the second phase was characterized by plant height stability in the period between male flowering and the end of the cycle.

There were no significant differences in plant height between the different plant densities, with an average height of 242 cm (Figure 3 and Appendix 3).

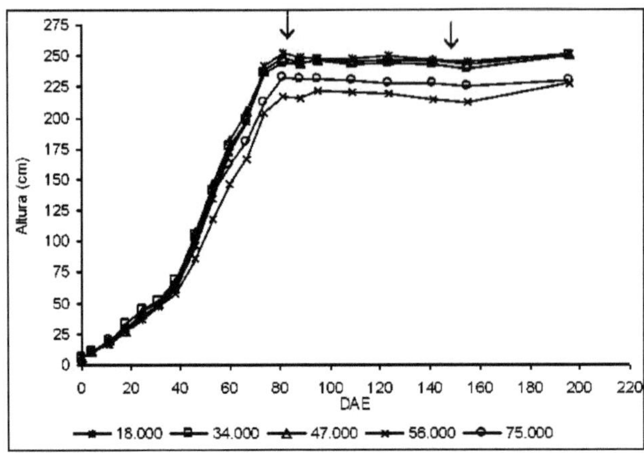

Figure 3. Plant height as a function of the number of days after emergence for the MPA1 maize variety. The arrows indicate male flowering and physiological maturity.

There were no differences in the ear insertion height of plants grown at different plant densities, with the average ear insertion height being 129 cm (Figure 4 and Appendix 4).

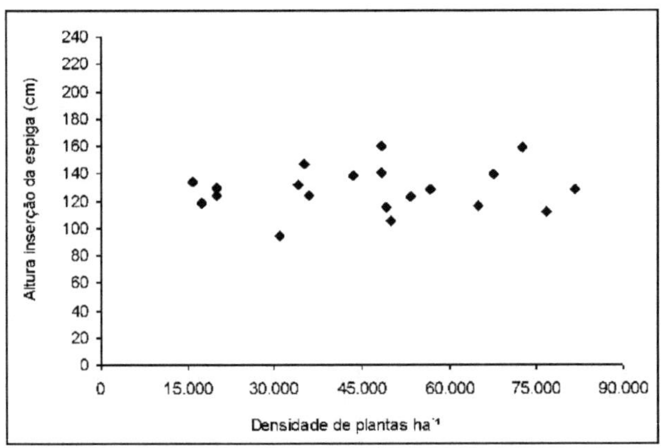

Figure 4: Ear insertion height as a function of plant density for the MPA1 maize variety.

4.2.2 Stalk diameter

During the cultivation of the MPA1 variety, the behaviour of the diameter of the stalks was characterized by two distinct phases of development, independent of the plant densities. The first phase was characterized by almost linear growth in diameter in the period between emergence and 53 DAE, while the second phase was characterized by a small reduction in the diameter of the stalks followed by stabilization until the end of the cycle (Figure 5).

Stalk diameter from 38 DAE onwards was significantly affected by the effect of plant densities (Appendix 5). There was a significant downward trend in diameter as plant densities increased (Figure 6). Close to flowering (81 DAE), populations containing 18,000 plants ha^{-1} had an average diameter of 32.8 mm, while populations with 75,000 plants ha^{-1} had an average diameter of 20.8 mm (Figure 5).

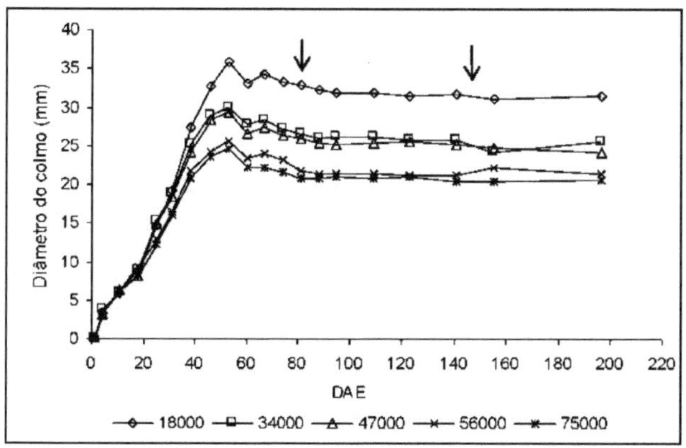

Figura 5. Stalk diameter as a function of days after emergence for maize variety MPA1. The arrows indicate male flowering and physiological maturity.

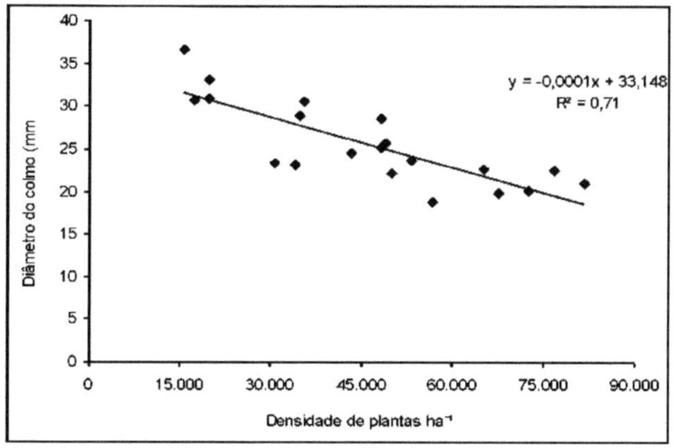

Figura 6. Stalk diameter at 109 DAE as a function of plant density for the MPA1 corn variety.

4.2.3 Number of expanded leaves

The rate of leaf emergence showed linear growth until 74 DAE (Figure 7). The total number of expanded leaves showed significant differences between the densities. This difference began at 31 DAE and continued until the end of the cycle (Appendix 6). There was also a significant downward trend in the number of leaves (22.5 to 21.1 leaves) as plant density increased from 18,000 to 75,000 plants ha^{-1} , respectively (Figure 8).

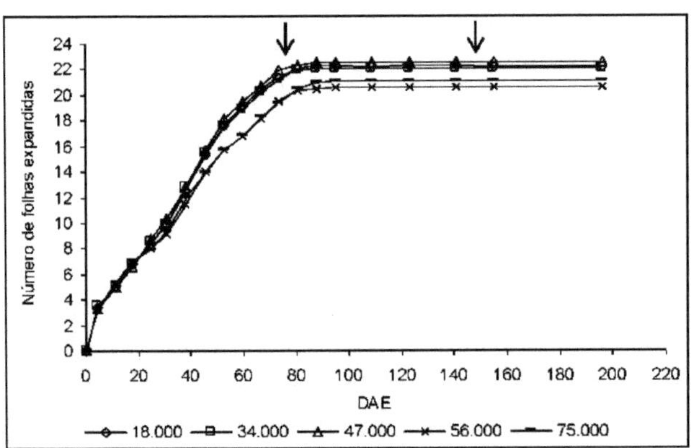

Figure 7. Number of expanded leaves as a function of the number of days after emergence for the MPA1 maize variety. The arrows indicate male flowering and physiological maturity.

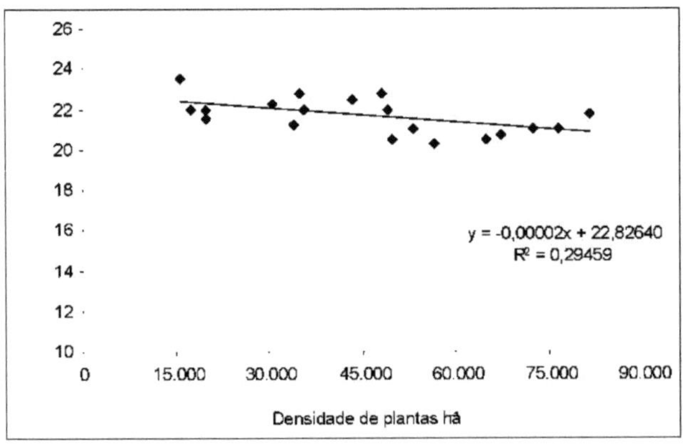

Figure 8. Total number of expanded leaves at 95 DAE as a function of plant density for

maize variety MPA1.

4.2.4 Leaf angle

Plant densities had no significant effect on the leaf angle of the first leaf above the main ear and the 12th leaf. The average leaf angle of the first leaf above the main ear was 23 degrees (Figure 9A and Appendix 7) while that of the 12th leaf was 18.7 degrees (Figure 9B and Appendix 7). However, there was a non-significant tendency for the leaf angle to decrease as plant density increased, especially for the 12th leaf (Figure 9B).

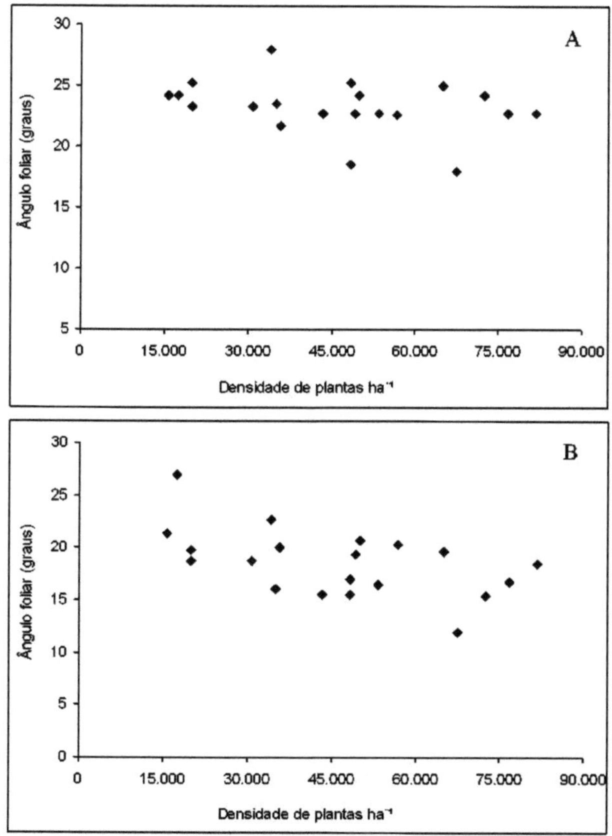

Figure 9. Leaf angle of the first leaf above the main ear (A) and leaf angle of the 12th leaf (B) as a function of plant density for the MPA1 maize variety.

4.2.5 Leaf area

Individual leaf area was significantly affected by the effect of plant density between the 16th and 20th leaf (Figure 10 and Appendix 8). There was a significant downward trend in individual leaf area as densities increased (Figure 11).

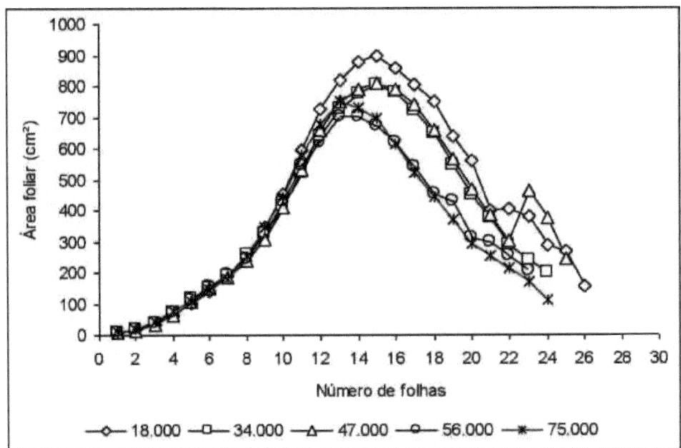

Figure 10. Individual leaf area in relation to the number of leaves from the base to the top of the vegetative canopy for different plant densities of the MPA1 corn variety.

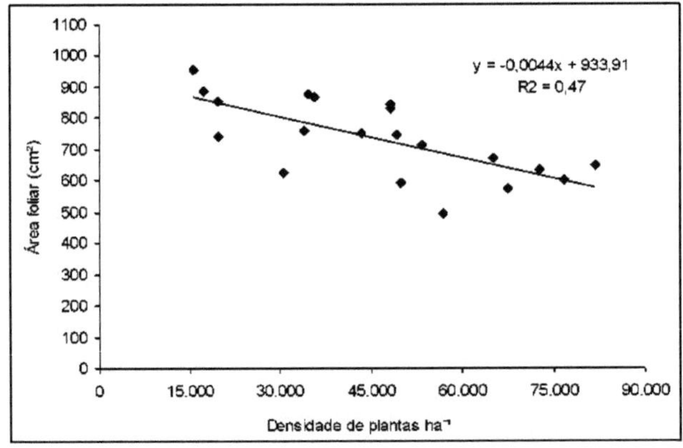

Figure 11. Individual leaf area of the 16th leaf as a function of plant density for the MPA1 maize variety.

The development of the total leaf area per plant showed similar behavior for the five densities. This variable was characterized by slow growth between emergence and 31 DAE, followed by an increase in this variable until close to male flowering (81 DAE) (Figure 12).

The total leaf area per plant showed significant differences between the plant densities from 53 DAE onwards, indicating the start of intraspecific competition which lasted until the end of the cycle (Appendix 9). There was a significant downward trend in the total leaf area per plant as the densities increased (Figure 13).

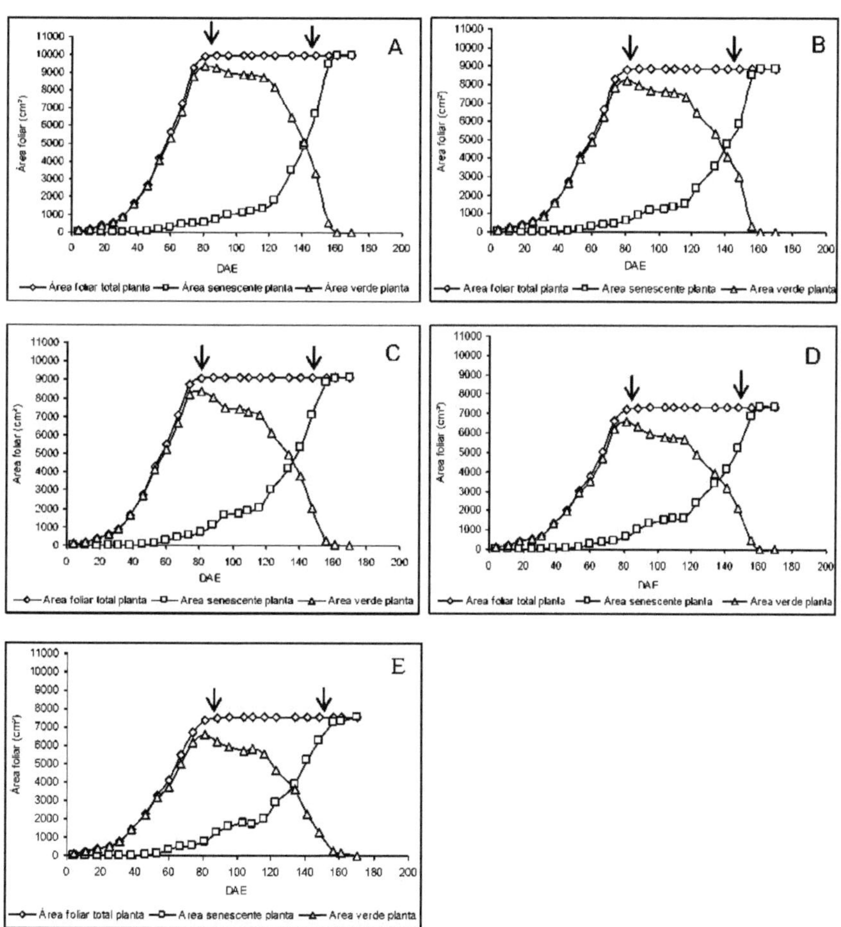

Figure 12. Total, green and senescent leaf area per plant as a function of days after emergence at densities of 18,000(A), 34,000(B), 47,000(C), 56,000(D) and 75,000(E) plants

ha^{-1} of the MPA1 maize variety. The arrows indicate male flowering and physiological maturation.

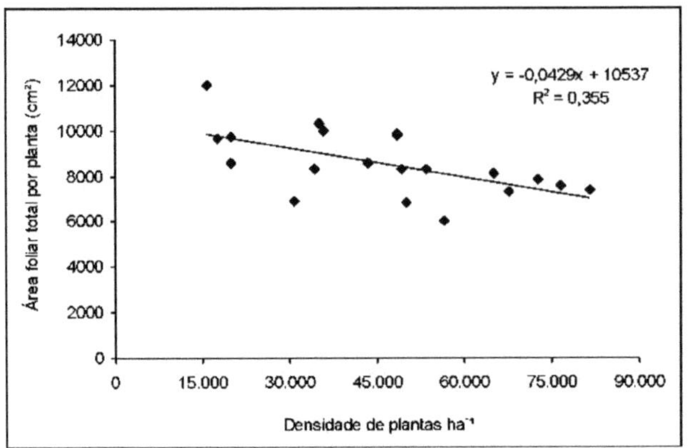

Figure 13. Total leaf area per plant at 95 DAE as a function of plant density for maize variety MPA1.

In general, for all five densities, the green leaf area per plant throughout the growing season was characterized by four distinct phases (Figure 12). The first phase showed a slow growth of green leaf area between emergence and 31 DAE, while the second phase was characterized by an almost linear development of green area per plant until male flowering at 81 DAE. In the third phase there was a stabilization of the green area followed by a small drop, while in the fourth phase the green area fell sharply.

Significant differences between the green leaf area per plant were identified for the plant densities from 53 DAE to 134 DAE (Appendix 10). There was a significant downward trend in green leaf area per plant as plant density increased, with the greatest differences identified close to male flowering (Figure 14).

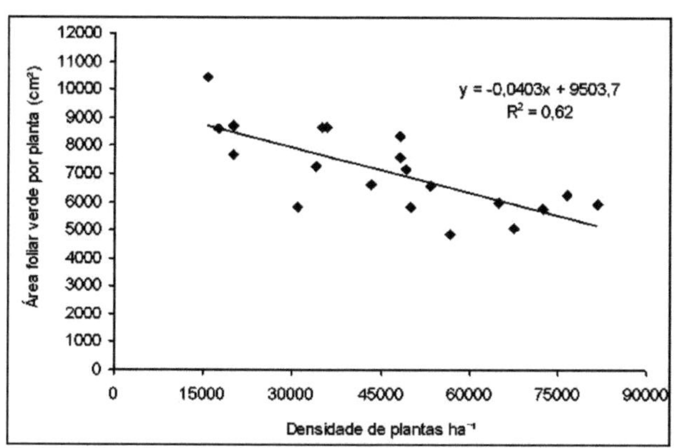

Figure 14. Green leaf area per plant at 104 DAE (near flowering) as a function of plant density for the MPA1 maize variety.

The senescent leaf area per plant, regardless of plant density, showed three distinct phases of development during the cycle of the MPA1 variety (Figure 15). The first phase occurred between emergence and the 30th DAE and was characterized by the absence of senescent leaf area during this period. The second phase was characterized by the start of leaf senescence at 31 DAE with an increase in this senescent area until approximately 116 DAE. The third phase was characterized by an accelerated increase in senescent leaf area per plant until the end of the crop (Figure 12).

The relative senescent leaf area per plant showed significant differences between the plant densities from 67 DAE onwards and remained so until 123 DAE (Figure 15 and Appendix 11). During this period, there was a significant tendency for the relative senescent area per plant to increase with the increase in population (Figure 16).

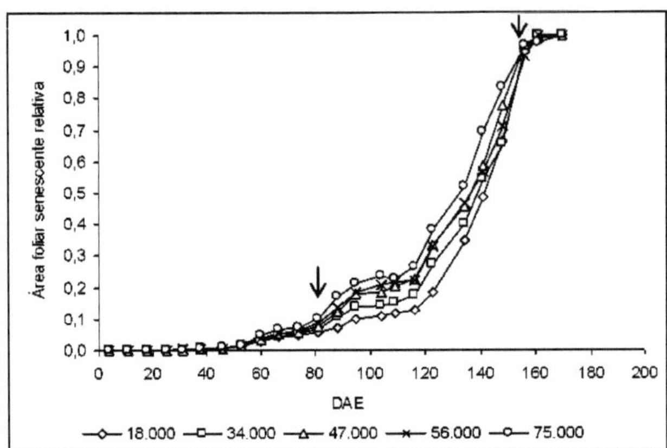

Figure 15. Relative senescent leaf area per plant as a function of days after emergence for maize variety MPA1. The arrows indicate male flowering and physiological maturity, respectively.

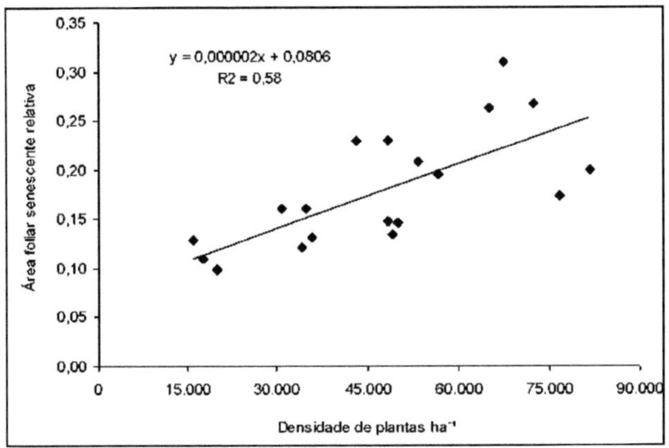

Figure 16. Relative senescent leaf area per plant as a function of plant density at 104 DAE for maize variety MPA1.

The leaf area index of the plant densities was characterized by four phases of development over the course of the crop (Figure 17). The first phase showed slow growth between emergence and 31 DAE, while the second phase, which began after this period, showed an almost linear increase until 81 DAE, when the highest indices were obtained. In the third phase, the leaf area index stabilized, especially at the lower densities, followed by a small drop until 116 DAE, while in the fourth phase there was a sharp decrease in the leaf area index, especially at the higher densities.

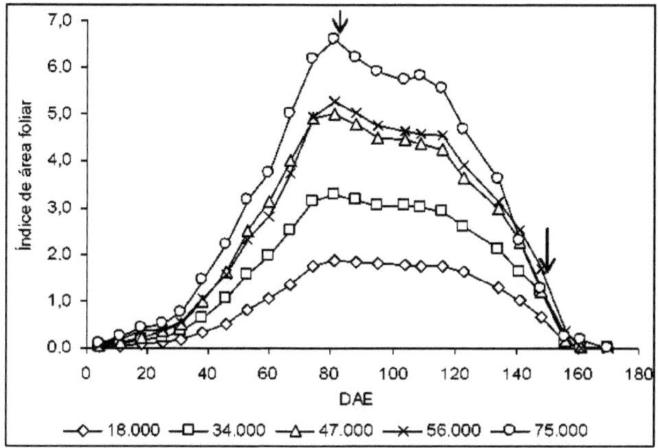

Figure 17. Leaf area index as a function of days after emergence for the MPA1 maize variety. The arrows indicate male flowering and physiological maturity.

The leaf area index of the plant densities showed significant differences between the treatments studied. These differences began at 4 DAE and continued until 134 DAE (Appendix 12), with a significant tendency for the leaf area index to increase as the plant density increased (Figure 18).

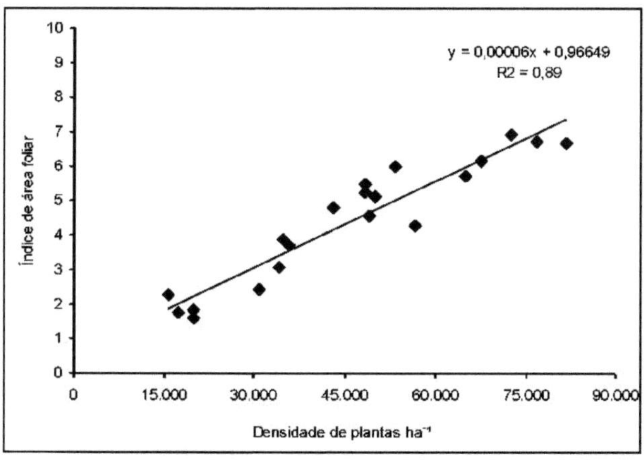

Figure 18. Leaf area index as a function of plant density at 79 DAE for maize variety MPA1.

4.3 Agronomic results

4.3.1 Total phytomass

The phytomass produced by the MPA1 variety differed between plant densities over the course of the crop (Figure 19). However, the total phytomass was determined at the leitoso grain phenological stage at 112 DAE due to the fact that this stage is close to the hard pasty grain stage which, according to Souza (1989), is the ideal point for harvesting corn for silage production due to the quantity of phytomass produced and the quality of the phytomass.

The total phytomass produced at the milky grain stage showed significant differences between the plant densities used. However, according to the Duncam 5% mean separation test, the treatments with 47,000, 56,000 and 75,000 plants ha^{-1} showed no significant differences between them (Table 5). With the regression analysis, it was possible to establish the optimum density of 57,000 plants ha^{-1} for the MPA1 variety, which achieved 15,725 kg ha^{-1} of phytomass (Figure 20).

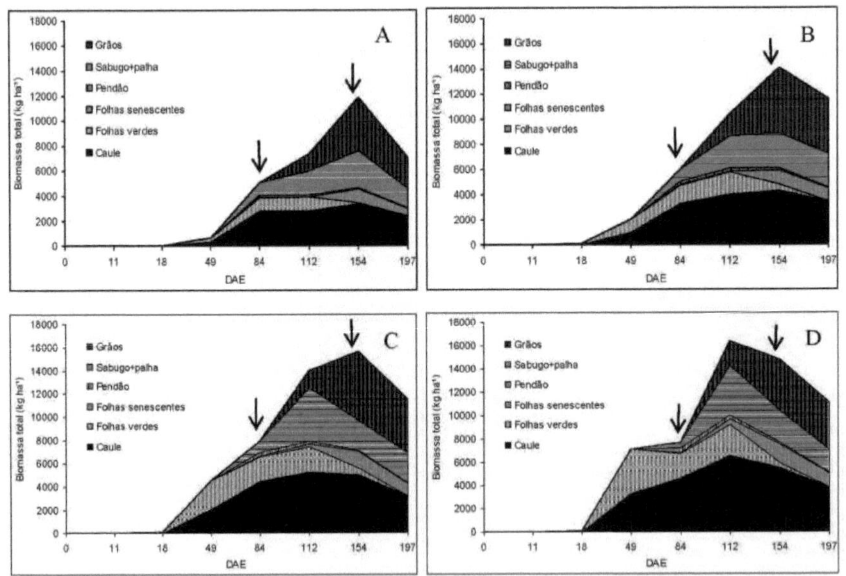

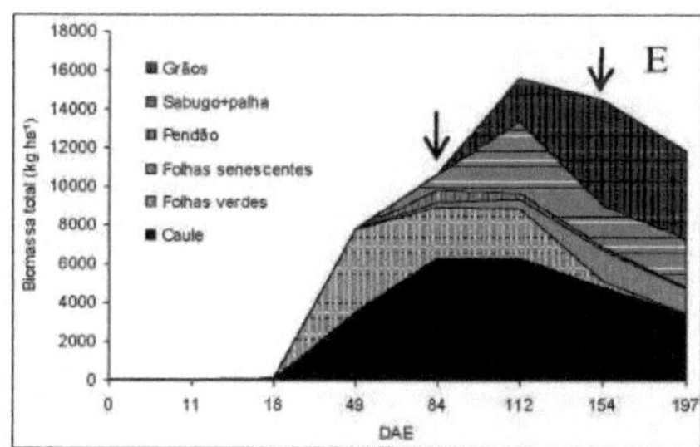

Figure 19. Total phytomass (stem, green leaves, dry leaves, stalk, cob and kernels) for the density of 18,000 (A), 34,000 (B), 47,000 (C), 56,000 (D) and 75,000 (E) plants ha- 1 as a function of the days after emergence of the MPA1 maize variety. The arrows indicate male flowering and physiological maturity.

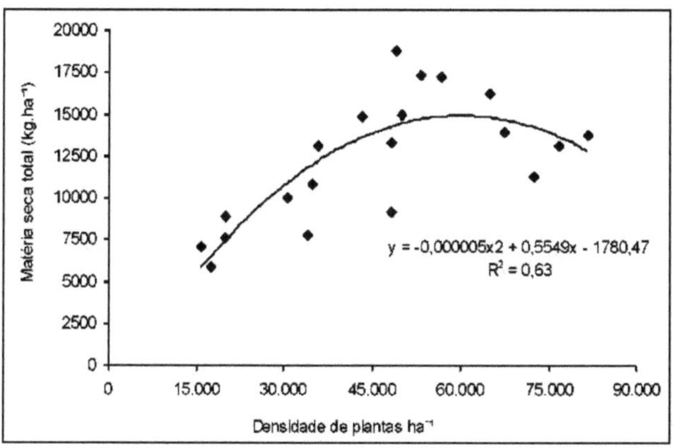

Figure 20. Total phytomass as a function of plant density at 112 DAE for maize variety MPA1.

The proportion of stems per plant showed no significant differences between the treatments used during the cycle (Appendix 15), but there were two distinct phases of development regardless of the treatments. The first phase was characterized by an increase in the proportion of stems per plant between the beginning of the development of the individuals and the male flowering of the variety, where this proportion represented 57% of the total phytomass of the plants. In the second phase there was a decrease in the proportion of stems (after male flowering) which continued until the material was harvested. During the physiological maturation of the plants, the phytomass of the stems represented 31% of the total phytomass of the plants (Figure 21).

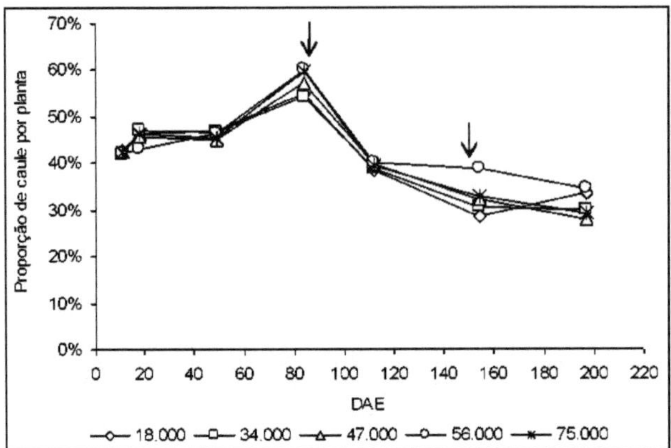

Figure 21. Proportion of stems per plant as a function of days after emergence for maize variety MPA1. The arrows indicate male flowering and physiological maturity.

The plant densities had no significant effect on the proportion of green leaves on the plants during the crop cycle (Figure 22 and Appendix 17). However, the proportion of these green leaves in relation to the total phytomass per plant varied over the course of the cycle, as initially the proportion of green leaves was 58% of the total phytomass of the plants, with a gradual reduction in the proportion over the course of the crop. At the male flowering stage of the variety, this proportion was reduced to 25% of the total plant phytomass. The reduction in the proportion of green leaves in relation to total phytomass continued until the end of the crop.

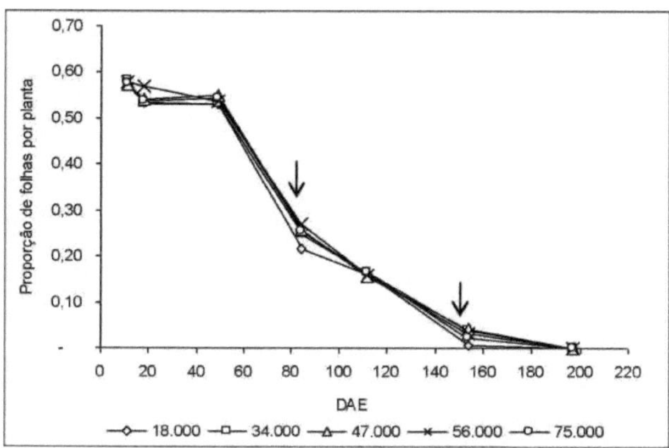

Figura 22. Proportion of green leaves per plant as a function of days after emergence for maize variety MPA1. The arrows indicate male flowering and physiological maturity.

Plant density had no significant effect on the harvest index of the MPA1 variety (Appendix 20). For the five densities used, the average grain yield obtained was equivalent to 38% of the total phytomass produced by the populations, generating a harvest index of 0.38 (Figure 23).

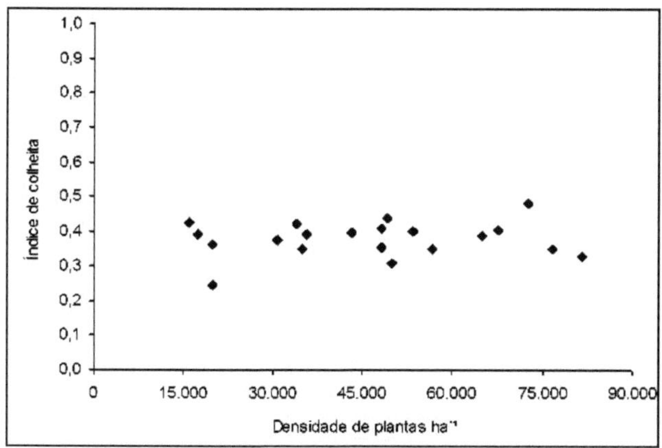

Figura 23. Harvest index as a function of plant density for the MPA1 corn variety.

4.4 Plant density Grain yield

4.4.1 Grain yield

The grain yield of the MPA1 variety showed significant results between the plant densities studied (Appendix 20). However, the treatments with 34,000, 47,000, 56,000 and 75,000 plants ha^{-1} did not differ significantly from each other, but were higher than the yield of the treatment with 18,000 plants ha^{-} ı (Table 6). Based on the regression analysis, the optimum density of 57,500 plants ha^{-} 1 was established for the MPA1 variety, which achieved a yield of 6,280 kg ha^{-} 1 (Figure 24).

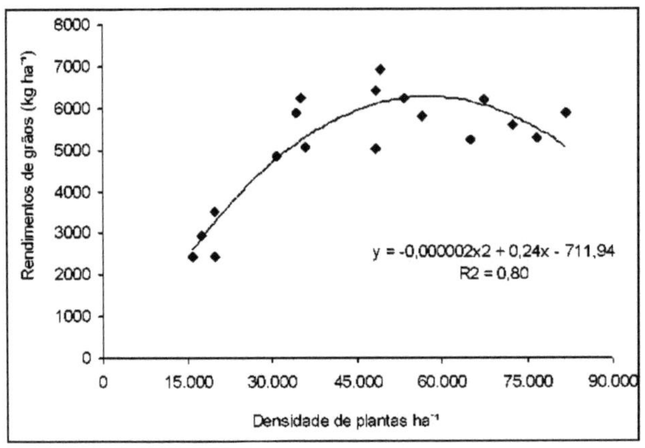

Figure 24. Grain yield as a function of plant density for maize variety MPA1.

The number of grains per unit area at harvest (197 DAE) showed a quadratic response to the plant densities studied (Appendix 20). There was an increase in the number of grains per unit area as densities increased up to 60,600 plants ha^{-1} , where 1,765 grains were obtained m^{-2} (Figure 25).

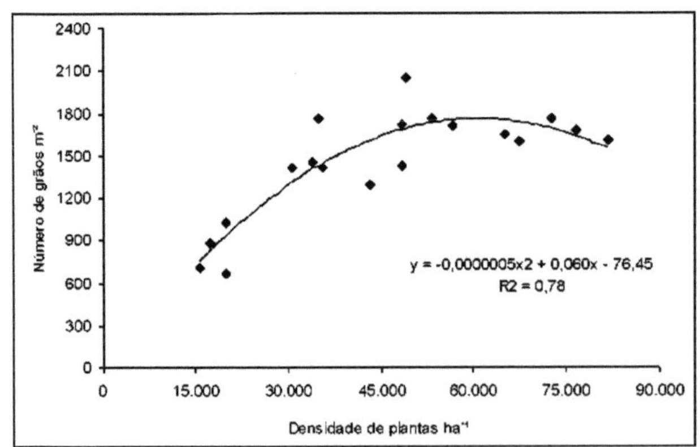

Figure 25. Number of kernels m^{-2} as a function of plant density at 197 DAE for maize variety MPA1.

The number of kernels on the cob^{-1} showed significant differences between the treatments. There was a downward trend in the number of kernels on the cob^{-1} as the densities increased (Figure 26 and Appendix 20).

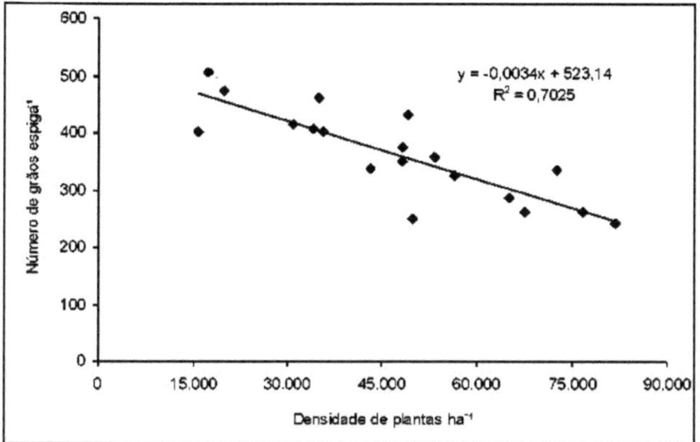

Figure 26. Number of kernels on the cob^{-1} as a function of plant density at 197 DAE for maize variety MPA1.

There was a significant increase in the number of spikes m^{-2} as plant densities increased (Figure 27 and Appendix 20).

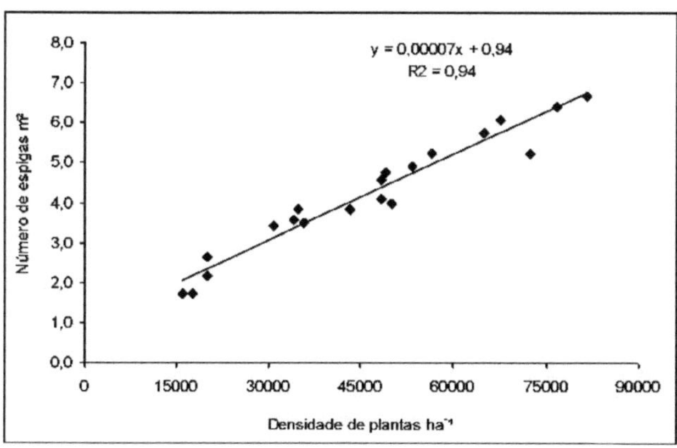

Figure 27. Number of ears m^{-2} as a function of plant density at 197 DAE for maize variety MPA1.

The number of spikes^{-1} decreased significantly with increasing plant density at 197 DAE (Figure 28 and Appendix 20).

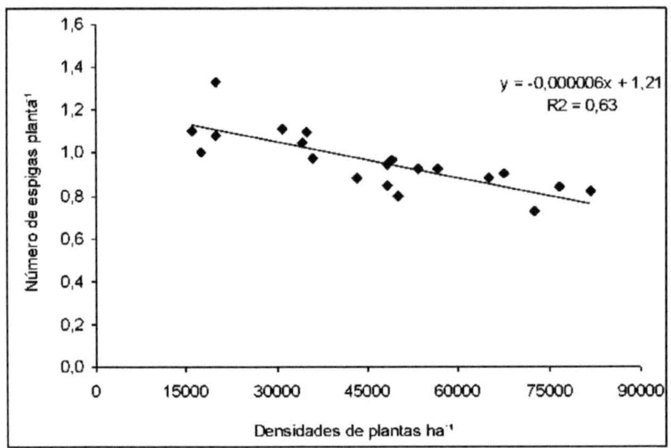

Figure 28. Number of ears^{M} as a function of plant density at 197 DAE for the MPA1 corn variety.

The weight of 1000 grains at 197 DAE showed significant differences between the plant densities studied (Appendix 20). A significant downward trend in the weight of 1000 grains was identified as plant densities increased.

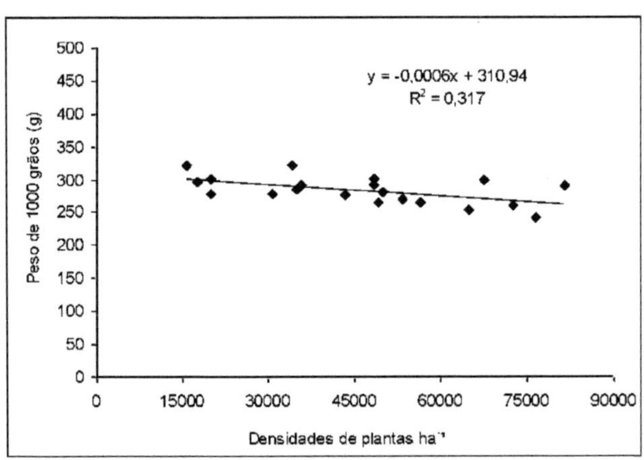

Figure 29. Weight of 1000 grains as a function of plant density at 197 DAE for maize variety MPA1.

4.5 Spontaneous plants

There was a significant reduction in the phytomass of spontaneous plants as plant density increased (Figure 30 and Appendix 21). This reduction in phytomass occurred approximately up to a density of 65,000 plants ha^{-1} . From this optimum density there was no significant reduction in phytomass.

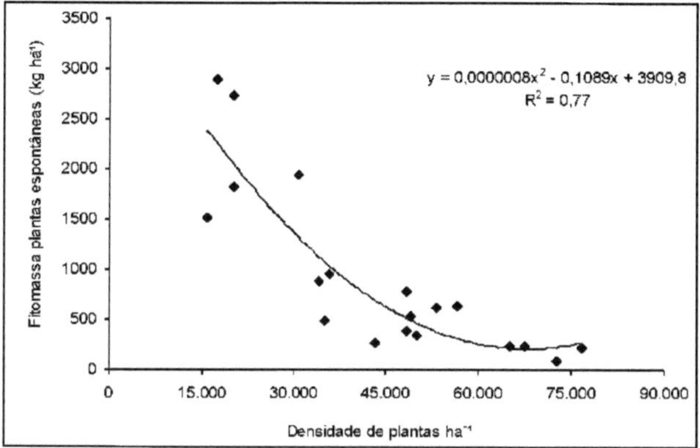

Figure 30. Phytomass of weeds as a function of plant density at 139 DAE for maize variety MPA1.

4.6 Plant lodging and breakage

Plant densities had a significant effect on lodging and plant breakage at 155 and 197 DAE (Appendix 22). There was a significant upward trend in lodging and plant breakage as densities increased (Figure 31 A and 31 B),

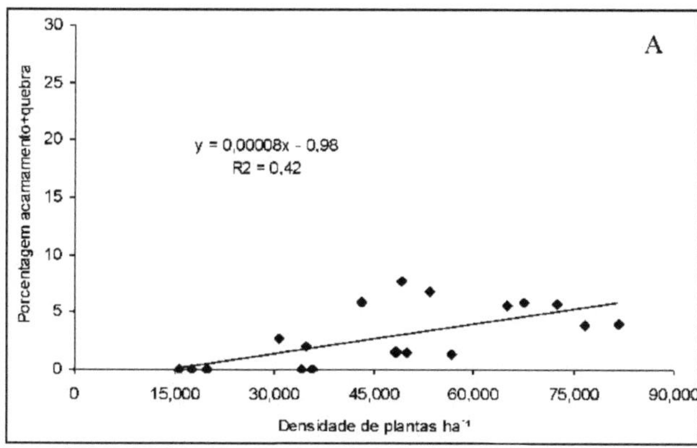

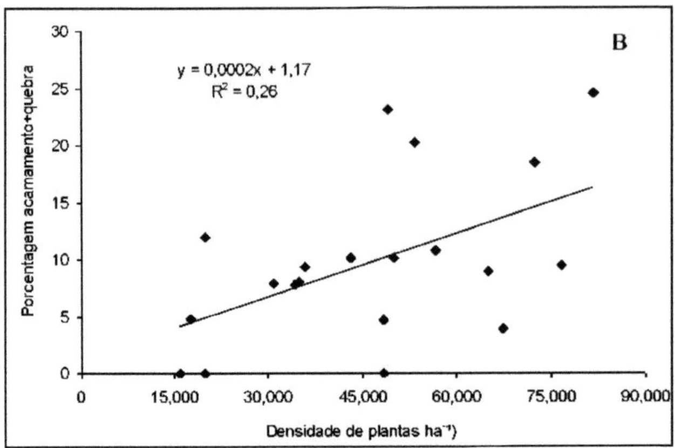

Figure 31. Percentage of lodging and plant breakage as a function of plant density at 155 (A) and 197 (B) DAE for maize variety MPA1.

5. Discussion

5.1 Intraspecific competition in plants

The increase in plant density led to an increase in intraspecific competition. This increase in competition was responsible for promoting a significant reduction in the amount of phytomass produced per plant during the crop cycle between the densities. However, although the increase in population caused changes in plant phytomass, the densities did not influence the partitioning of this phytomass between the organs of the plant, so much so that the proportion of stems per plant (Figure 21), the proportion of leaves per plant (Figure 22) and the variety's harvest index (Figure 23) remained unchanged.

This homogeneity obtained between the harvest indices was possibly due to the fact that the effects of the densities tested were insufficient to cause changes in this variable, since Hashemi et al. (2005), using densities between 90,000 and 120,000 plants ha^{-1} , identified a significant reduction in the harvest index as the population increased. The lack of effects of intraspecific competition at the densities tested on the harvest index is probably related to the fact that these densities did not promote changes in the growth rates of the populations, because according to Pagano and Maddonni (2007), changes in growth rates as a result of intraspecific competition result in changes in the partitioning of phytomass and consequently changes in harvest indices. In this way, the intra-specific competition caused by the densities significantly influenced the phytomass produced per plant, but did not influence the partitioning of the phytomass produced between the different organs of the plant.

The alteration caused to the phytomass produced by the plants is due to the fact that plant densities increase intraspecific competition, causing environmental impoverishment in terms of the resources available per plant, thus compromising the growth and development of the individuals (Maddonni & Otegui, 2006). In addition to promoting a decline in plant phytomass

(Maddonni & Otegui, 2006), intraspecific competition is also responsible for causing some morphophysiological changes in plants (Cruz et al., 2006a).

Some morphophysiological changes caused by the increase in densities were identified. Among the morphophysiological variables that were reduced by the increase in competition were the number of expanded leaves per plant (Figures 7 and 8), the diameter of the thatch (Figures 5 and 6), the total leaf area per plant (Figure 13), the individual leaf area (Figures 10 and 11) and the green leaf area per plant (Figure 12). Similar results caused by increased intraspecific competition were also found by Modarres et al. (1998) for the number of expanded leaves; Palhares (2003) and Sangoi et al. (2001a) for thatch diameter; Silva et al. (1999) for total leaf area per plant; Sangoi et al. (2002c), Maddonni et al. (2001) and Modarres et al. (1998) for individual leaf area and Borrâs et al. (2003), Maddonni et al. (2001) and Durâes et al. (1995) for green leaf area per plant.

The relative senescent area per plant also changed with the increase in population, and these changes led to an increase in this variable over the course of the crop (Figure 15), in the same way as observed by Borràs et al. (2003) and Duraes et al. (1995). Although the increased plant densities were responsible for some morphophysiological changes in the plants, they were not able to promote changes in leaf angle (Figure 9), plant height (Figure 3) and ear insertion height (Figure 4).

5.2 Optimum plant density for silage production

In Santa Catarina, it is estimated that around 50,000 producers are involved in dairy farming (Marcondes, 2004). However, of the total produced by these farmers, around 70% comes from rural properties in western Santa Catarina (Sintese... 2005), where silage during the winter period becomes a good alternative food for dairy cattle.

The MPA1 variety used in this study is one of the varieties used for silage production in the far west of Santa Catarina. The amount of phytomass produced by this variety (Figure 20) showed quadratic behavior in relation to the densities tested, initially showing an increase in yields as a result of increasing densities, up to the optimum density of 57,500 plants ha^{-1} . This initial increase in yields caused by raising the densities to the optimum cultivation density was also observed by Widdicombe & Thelen (2002), Cox (1996) and Hashemi et al. (2005) who obtained optimum densities higher than the one obtained in this study (88,900, 90,000 and 90,000 plants ha^{-1} respectively) due to the fact that they used improved varieties in their work.

This behavior of the MPA1 variety resulted from the fact that the densities increased the leaf area index (Figures 17 and 18), as also observed by Silva et al. (1999), Westgate et al. (1997), Cox, (1996) and Tollennar & Bruulsema, (1988), providing the populations with an increase in the interception of the radiation incident on the crops and consequently in the transformation of this radiation into phytomass. Almeida et al. (2003) reported a positive correlation between phytomass and leaf area index, while Romano (2005) identified a linear increase in phytomass production caused by an increase in the interception of photosynthetically active radiation in crops.

When densities higher than the optimum crop density were evaluated, there was a drop in phytomass production, even though the supra-optimum densities had higher leaf area indices. This means that by increasing plant densities above the established optimum density, the radiation interception surface was increased, but the efficiency with which the plants converted this radiation into phytomass was reduced, resulting in under-utilization of this radiation. Manteith (1977) described that the phytomass produced by maize is the result of the radiation intercepted by the plants and the efficiency with which this radiation is used. Bearing in mind that the radiation intercepted by the MPA1 variety increased as a result of the higher leaf area index and phytomass production decreased, it is likely that the efficiency of using this radiation to convert it into phytomass was reduced. Tollenaar and Bruulsema (1988) confirmed this reduction, as in their studies there was a significant increase in the photosynthetically active radiation absorbed by the plants as plant densities increased (60,000 to 100,000 plants ha^{-1}). However, the data on the efficiency of converting the absorbed radiation into phytomass suggests small differences in the reduction caused by increasing the densities used, demonstrating that the greater amount of radiation intercepted/absorbed is not necessarily converted into phytomass (Tollenaar & Bruulsema 1988). The reduction in the efficiency of converting intercepted radiation into grain phytomass was also observed among the supra-optimal plant densities used by Andrade et al. (1993).

5.3 Plant density and grain yield

According to Sinclair (1998), maize grain yield can be expressed as a function of the phytomass produced and the variety's harvest index. Based on this information and the fact that the harvest index (Figure 23) of the MPA1 variety showed no significant differences between the plant densities used, it can be concluded that the increase in grain yield (Figure 24) occurred in response to the increase in phytomass at the higher densities (Figure 20), and not due to the partitioning of this phytomass (Figure 19). Therefore, the behavior of the variety associated with the increase in population increased the grain yield up to the optimum density of 57,000 plants ha^{-} 1 (Figure 24), as observed by Hashemi et al. (2005), Flesch & Vieira (2004), Borghi et al. (2004), Palhares (2003), Sangoi et al. (2002c), Sangoi & Salvador (1996) and Tollenaar & Bruulsema (1988) up to the optimum densities of 90,000, 73,955, 62,916, 60,000, 71,000, 75,000 and 100,000 plants ha^{-1} respectively. This increase in yield resulting from the increase in phytomass produced led to an increase in the number of grains per unit area (Figure 25), which was the yield component responsible for the increase in grain productivity. Massignam (2003); Sangoi et al. (2002c) and Cox (1996) also describe the number of grains as the yield component responsible for the increase in grain yield. Balbinot et al. (2005)

also report that, in addition to the number of grains per area, another relevant variable for determining grain yield in open-pollinated varieties is the number of grains per row of ears.

On the other hand, an increase in plant density above the optimum density determined a decrease in yield, possibly caused by a reduction in the number of grains per area (Figure 25) and the weight of the grains (Figure 29). According to Carcovà et al. (2002), this reduction in the number of grains per area is possibly associated with the increase in the interval between male and female flowering caused by the increase in plant density of the variety (Figure 1), as also observed by Tokatlidis & Koutroubas (2004), Sangoi et al. (2002a), Sangoi et al. (2001a) and Bolanos & Emeades (1996). This is because denser populations show a marked decrease in the growth rate of lateral buds in relation to the plant's apical growth point (shoot), leading to an increase in the time lag between the shoot and the ear (Sangoi, 2000, Sangoi & Salvador, 1996), compromising the plant's ability to grow. Salvador, 1996), compromising fertilization and reducing the number of fertilized spikelets, which leads to a reduction in the number of grains per area (Sangoi et al., 2001a, Sangoi & Salvador, 1998).

The reduction in the weight of 1000 grains caused by the increase in plant density, also observed by Maddonni & Otegui (2006), Flesch & Vieira (2004), Norwood (2001), Echarte et al. (2000) and Cox (1996), is probably associated with the reduction in the period between female flowering and the physiological maturity of the plants (Figure 2), as identified by Sangoi et al. (2002d). This period includes the grain filling phase, where a reduction in the number of days in this phase implies a reduction in grain weight (Sangoi et al., 2002a).

The percentage of lodged and broken plants showed a significant increase, as described by Borghi et al. (2004), Sangoi et al. (2001b), Sangoi et al. (2001a), caused by the increase in plant density (Figure 31), which can reduce the yield and quality of the grain. The increase in density led to an increase in intraspecific competition, which resulted in a decrease in the diameter of the stalk (Figure 6). According to Sangoi et al. (2002b) this differentiation in the diameter of the stalks resulted from plants grown in smaller populations intercepting greater amounts of individual solar radiation per leaf, promoting an increase in the photosynthetic capacity of the plants and generating greater availability of photoassimilates for the formation of structural compounds that minimize the fragility of the stalks, unlike denser populations. At higher densities, this availability of photoassimilates is lower, which is why smaller and more fragile stalks are formed. In addition to stalk diameter, other variables such as plant height (Figure 3) and ear insertion height (Figure 4) are also relevant for understanding lodging and plant breakage, due to the pendulum formed between the plant (ear) and the ground. However, for the MPA1 variety, these variables did not change between the densities used.

In order to obtain good grain and phytomass yields based on the agro-ecological cultivation system, the optimum density established is approximately 57,000 plants ha^{-1} . However, this condition is recommended for farmers who grow their crops on sites with high soil fertility and little water restriction, such as those used in this experiment, because any variation in fertility or water availability affects the optimum density (Cruz et al., 2006a). Therefore, in conditions where nutrients and water are limited, plant density should be reduced in order to minimize plant competition for these resources (Silva et al., 2006).

As well as the density of 57,000 plants ha- 1 being recommended for grain and phytomass production, it is also suitable for controlling spontaneous plants, as it is close to the ideal density for controlling spontaneous plants between rows (Figure 30). Using this density allows farmers to adopt a more economical, rational and environmentally friendly management technique (Balbinot & Fleck, 2004).

5.4 Correlation analysis

The analyses carried out between grain yield, dry phytomass, morphology and leaf area showed significant correlations (Appendix 23). Grain yield showed a positive correlation with dry phytomass (r=0.87), demonstrating that an increase in phytomass is associated with gains in yield, as identified by Tollenaar & Wu (1999); Sangoi & Salvador (1997) and Boukerrou & Rasmusson, (1990). According to Sangoi & Salvador (1997), the increase in phytomass is relevant for determining the increase in grain yield, however, another way of increasing grain yield is by increasing the harvest index of a variety (Sangoi & Salvador; Boukerrou & Rasmusson, 1990). The harvest index obtained by the variety under study showed no significant correlation with grain yield, so the increase in grain yield was due exclusively to the increase in phytomass.

The increase in phytomass in population density studies occurs in response to the increase in leaf area index (Silva *et al.* 1999; Westgate *et al.* 1997; Cox, 1996 and Tollennar & Bruulsema, 1988). Increasing the leaf area index maximizes the interception of solar radiation (Almeida et al., 2003), so that this increase generates the greatest accumulation of dry phytomass per area (Silva et al., 2006). Determining the leaf area index of a population depends on the total leaf area per plant and plant density (Tollennar & Wu, 1999). The total leaf area per plant of the MPA1 variety showed no correlation with the leaf area index, indicating that the increase in leaf area index was solely a function of the increase in plant density.

Like phytomass, the increase in grain yield showed a positive correlation with the leaf area index (r= 0.64). The increase in leaf area index (increased interception of solar radiation) led to an increase in the number of ears per area (r= 0.89) and the number of kernels per area (r= 0.68). Tollennar & Wu (1999) describe the number of grains per area and the number of ears per area as the most important variables in determining grain yield. The number of ears per plant (TOLLENNAR & WU 1999) and grain weight are also presented as relevant variables for determining yield, however, these variables showed a negative correlation with the leaf area index (r= -0.85 and -0.50, respectively). This decrease in the variables caused by an increase in the leaf area index was also observed by Maddonni & Otegui (2006), Flesch & Vieira (2004), Norwood (2001), Echarte et al. (2000) and Cox (1996) and is caused by an increase in intraspecific competition (Loomis & Connor 1992) at high plant densities (Silva et al., 2006). Therefore, as the leaf area index increased, the grain yield increased due to the increase in the number of ears per area and the number of kernels per area. On the other hand, as the leaf area index increased, the number of ears per plant and the weight of kernels decreased, but on balance there was an increase in the grain yield.

Grain yield showed a positive correlation with the percentage of lodging and plant breakage (r= 0.48). This relationship is due to the fact that the increase in the amount of phytomass per area and the leaf area index show negative correlations with the diameter of the thatch (r=-0.58 and -0.71 respectively). The effect of reducing the diameter of the thatch in individuals grown at high densities occurs because in these conditions of high competition the individuals preferentially invest their photosynthates in part of the area (mainly leaves), causing a reduction in the components destined for the structure of the thatch, leading to changes in the diameter and strength of this organ (Durâes et al., 1995; Loomis & Connor 1992). Another relevant factor that contributes to the reduction in the diameter and strength of the thatch is the fact that, after flowering and filling the kernels, the plants migrate their accumulated reserves from the stem to the kernels (SILVA et al., 2006). This is a way of compensating for competition between plants, as the leaf area index is higher at high densities, while the leaf area per plant is lower. Thus, plants grown at high densities compensate for the demand for photosynthates by migrating reserves from the stems to the kernels, leading to an increase in lodging and plant breakage and a reduction in yields at the harvest stage.

6. Conclusion

- The use of different plant densities for the MPA1 variety led to an increase in intra-specific competition. As a result, there was a reduction in phytomass per plant as plant density increased, but plant density had no effect on the partitioning of phytomass between organs.

- Increased intraspecific competition at higher densities led to morphophysiological changes in the plants, such as a reduction in the number of leaves, the diameter of the stalk, the total leaf area per plant, the individual leaf area, the green leaf area per plant and an increase in the relative senescent area per plant.

- Intraspecific competition caused changes in plant phenology, increasing the sub-period between emergence and male and female flowering of the variety, increasing the interval between flowering and reducing the period between female flowering and physiological maturity.

- The optimum density for producing grain and phytomass for silage was 57,000 plants ha^{-1} , and the variation in grain yield as a function of plant density was mainly due to the variation in phytomass, as the harvest index had no significant effect between plant densities.

7. Bibliographical references

ALVES, A. C.; VOGT, G. C.; FANTINI, A. C.; OGLIARI, J. B.; MARASCHIN, M. **Local maize varieties and family farming in the far west of Santa Catarina**. In. CANCI, A., VOGT, G. A., CANCI, I. J. A diversidade das espécies crilas em Anchieta - SC. Sâo Miguel do Oeste, 2004a. p.67-76.

ALVES, A. C.; OGLIARI, J. B.; VOGT, G. C.; CANCI, A.; BALBINOT, A. A. Jr., FONSECA, J. A. **Criollo maize: grain yield and agronomic characteristics.** In. CANCI et al. 2004. The diversity of Creole species in Anchieta - SC. Sâo Miguel do Oeste, 2004 b. p.87-94.

ALMEIDA, M. L.; SANGOI, L.; NAVA, I. C.; GALIO, J.; TRENTIN, P. S.; RAMPAZZO, C. Maize initial growth and its relationship with grain yield. **Ciência Rural**. v. 33, p.189-194, 2003.

ALMEIDA, M.L.; MEROTTO, A.; SANGOI, L. Increasing plant density: an alternative for increasing corn grain yield in regions with a short summer growing season. **Ciência Rural**. v.30, n.1, p.23-29, 2000.

ANDRADE, F. H.; UHART, S. A.; FRUGONE, M. I. Intercepted radiation at flowering and kernel number in maize: Shade versus plant density effects. **Crop Science**. v.33, p.482-485, 1993.

ARGENTA, G.; SILVA, P. R. F.; SANGOI, L. Plant arrangement in corn: Analysis of the state of the art. **Ciência Rural**. v.31, n.6, p.10751084, 2001.

BALBINOT, A. A. Jr.; ALVES, A. C.; FONSECA, F. J. da. Plant density in open-pollinated maize varieties under agroecological cultivation. 5th Santa Catarina Technical Meeting on Maize and Beans. Expanded abstract. Chapecó, 2005, p.82-86.

BALBINOT, A. A. Jr.; FLECK, N. G. Weed management in corn crops as a function of plant spatial arrangement and genotype characteristics. **Ciência Rural**, v.34, n.6, p.245-252, 2004.

BALBINOT, JR, A.; BACKES, R. L.; ALVES, A. C.; OGLIARI, J. B.; FONSECA,

J. A. Contribution of yield components to grain yield in open-pollinated maize varieties. **Revista brasileira de agrociência**. v.11, n. 2, p. 161-166, 2005.

BALBINOT, A. A. Jr.; FONSECA, F. J. da.; ALVES, A. C.; OGLIARI, J. B.; FANTINI, A. C.; MARASCHIN, M. Plant characteristics, lodging and stalk breakage in open-pollinated maize varieties. Santa Catarina maize and beans technical meeting. Expanded abstract. Lages, CAV UDESC, 2003, p.160-164.

BOLANOS, J.; EMEADES, G. O. The importance of the anthesissilking interval in breeding for drought tolerance in tropical maize. **Field Crops Research**. v.48, p. 65-80, 1996.

BORGHI, E.; MELLO, L. M.; CRUSCIOL, C. A. C. Area and plant fertilization, population density and maize development as a function of soil management system. **Acta Scientiarum Agronomy**. v.26, n.3, p.337-345, 2004.

BORRAS, L.; MADDONNI, G. A.; OTEGUI, M. E. Leaf senescence in maize hybrids: plant population, row spacing and kernel set effects. **Field Crops Research**. v.82, p.13-26, 2003.

BOUKERROU, L.; RASMUSSON, D. C.; Breeding for high biomass yield in spring Barley. **Crop Science**. Madison. V. 30, p.31-35, 1990.

CANCI, A.; VOGT, G. A.; CANCI, I. J. **The diversity of Creole species in Anchieta - Sc**. Sâo Miguel do D'Oeste, 2004. 112p.

CANCI, A. **Creole seeds: Building sovereignty at the farmer's hand**. Sâo Miguel D'Oeste, 2002.160p.

AGRICULTURAL CENSUS 1995/1996. **Establishments by total area group. Santa Catarina, Brazil**. Available at: www.ibge.gov.br. Accessed on: October 10, 2005.

CIASC. **Santa Catarina Information and Automation Center**. Available at: www.mapainterativo.ciasc.gov.br/sc. Accessed on: September 14, 2005.

CLEMENTS, H. F. **Interactions of factors affecting yield**. Hawaii Agricultural

Experiment Station. n.653, p.409-442, 1963.

CONAB. **National Supply Company**. Available at: www.conab.gov.br. Accessed on: September 15, 2005.

COSTA, A S. V. da.; GALVAO, E. R.; SILVA, M. B.; PREZOTTI, L.; RIBEIRO, J. M. O. Population densities of maize in the Vale do Rio Doce region. **Revista Ceres**. v.52, p.33-34, 2005.

COX, W.J. Whole-plant physiological and yield responses of maize to plant density. **Agronomy Journal**. v.88, p.489-496, 1996.

CRUZ, J. C.; PEREIRA, I. A.F.; ALVARENGA, R. C.; GONTIJO, M. M. N.; VIANA, J. H. M.; OLIVEIRA, M. F. de.; SANTANA, D. P. Corn crop management. Embrapa milho e sorgo. Sete Lagoas MG. EMBRAPA. **Technical circular** n.87, p.12, 2006a.

CRUZ, F. T da; BATISTA, M. A.; MONTEIRO, F. RIEG, F.; SILVA, J. da; HEINZEN, T. SCHMIDT, L. Participatory agrobiodiversity management strategies in the municipality of Anitàpolis. In: KIST, V.; SANTOS, K. L. **Estratégias participativas de manejo da agrobiodiversidade**. Florianópolis-SC, 2006b, p.57-68.

DIDONET, A. C.; RODRIGUES, O.; MARIO, J. L.; IDE, F. Effects of solar radiation and temperature on the definition of kernel number in corn. **Pesquisa Agropecuâria Brasileira**. v.37, n.7, p.933-938, 2002.

DROUET, J. L.; MOULIA, B. Spatial re-orientation of maize leaves affected by initial plant orientation and density. **Agricultural and Forest Meteorology**. v.88, p.85-100, 1997.

DURAES, F. O. M., MAGALHAES, P. C., GAMA, E. E. G., OLIVEIRA, A. C. de. Phenotypic characterization of maize lines in terms of yield and photosynthetic efficiency. **Revista Brasileira de Milho e Sorgo**, v.4, n.3, p.355-361, 2005.

DURAES, F. O. M.; MAGALHAES, P. C.; COSTA, J. D.; FANCELLI, A. L.

Ecophysiological factors affecting the behavior of corn in late sowing (safrinha) in central Brazil. **Scientia Agricola**. v.52, p.491-501, 1995.

DUVICK, D. N.; CASSMAN, K. G. Post-Green Revolution Trends in Yield Potential of Temperate Maize in the North-Central United States. **Crop Science**. V.39, p.1622-1630, 1999.

ECHARTE, L.; LUQUE, S.; ANDRADE, F. H.; SADRAS, V. O.; CIRILO, A.; OTEGUI, M. E.; VEJA, C. R. C. Response of maize kernel number to plant density in Argentinean hybrids released between 1965 and 1993. **Field Crops Research**. v.68, p.1-8, 2000.

EMBRAPA CORN AND SORGHUM. **National Maize and Sorghum Research Center**. Available at: www.cnpms.embrapa.br. Accessed on: September 16, 2005.

FANCELLI, A. L.; DOURADO NETO, D. **Corn production**. Guaiba: Agropecuària, 360p. 2000.

FELLNER, M., HORTON, L. A., COCKE, A. E. STEPHENS, N. R. FORD, E. D., Light interacts with auxin during leaf elongation and leaf angle development in young corn seedlings. **Planta.** v.216, p.366-376, 2003.

FLESCH, R. D.; VIEIRA, L. C. Spacing and densities of corn with different cycles in western Santa Catarina, Brazil. **Ciência Rural**. v.34, p.23-51, 2004.

GARDIOL, J. M.; SERIO, L. A.; DELLA MAGGIORA, A. I. Modelling evapotranspiration of corn (Zea mays) under different plant densities. **Journal of Hydrology.** V.271, p.188-196, 2003.

GOULART, S.; GUGEL, G. Participatory agrobiodiversity management strategies in the municipality of palmitos. In: KIST, V.; SANTOS, K. L. **Estratégias participativas de manejo da agrobiodiversidade**. Florianópolis, 2006. p.31-39.

HARLAN, J. R. **Crops and Man**. American Society of Agronomy and Crop Science Society of America, Madison Wiscosin, 295p. 1975.

HASHEMI, A.; HEBERT, S. J.; PUTMAN, P. H. Yield response of corn to crowding stress. **Agronomy Journal**. v.97, p.839-846, 2005.

JACOBS, B. C.; PEARSON, C. J. Potential yield of maize, determined by rates of growth and development of ears. **Field Crops Research**. v.27, p.281-298, 1991.

JARVIS, D. I.; MYER, L.; KLEMICK, H.; GUARINO, L.; SMALE, M.; BROWN, A. D. H.; SADIKI, M.; SHAPIT, B.; HODGKIN, T. A **Training Guide for In situ Conservation On farm**. IPGRI, Rome, 193p, 2000.

KAMARA, A. Y.; MENKIR, FAKOREDE, M. A. B. AJALA, S. O.; BADU-APRAKU, B.; HUREH, I. Agronomic performance of maize cultivars representing three decades of breeding in the Guinea Savanas of West and Central Africa. **Journal of Agricultural Science**. v.142, p.567-575, 2004.

KINIRY, J. R.; KNIEVEL, D. P. Response of maize seed number to solar radiation intercepted soon after anthesis. **Agronomy Journal**. v.87, p.228-234, 1995.

KIST, V. **Recurrent selection of half-sib families in a composite maize population from Anchieta - SC**. 2006. 150p. Dissertation (Master's Degree in Plant Genetic Resources) - Federal University of Santa Catarina, Florianópolis, 2006a.

KIST, V.; GUEDINI, O.; CARLESSI, A.; RISSO, F. A. Participatory agrobiodiversity management strategies in the municipalities of Sâo Lourenço do Oeste, Novo Horizonte and Sâo Bernardinho. In: KIST, V.; SANTOS, K. L. **Estratégias participativas de manejo da agrobiodiversidade.** Florianópolis, 2006b. p.41-56.

KIST, V.; DALTÓE, E.; BRUNETTE, D. R. Participatory agrobiodiversity management strategies in the conquista da fronteira settlement, municipality of Dionisio Cerqueira. In: KIST, V.; SANTOS, K. L. **Estratégias participativas de manejo da agrobiodiversidade.** Florianópolis, 2006c. p.11-20.

LAUS NETO, J. A.; PUNDEK, M.; RIGO, P. H. POLA, A C. **Inventory of the lands of the Campos Novos Experimental Station**. Florianópolis. Epagri, 58p, 1999.

LINDQUIST, J. L.; ARKEBAUER, T. J.; WALTERS, D. T.; CASSMAN, K. G.; DOBERMANN, A. Maize radiation use efficiency under optimal growth conditions. **Agronomy journal**. v.97, p.72-78, 2005.

LOOMIS, R.S.; CONNOR, D. J. Crop ecology. **Productivity and management in agricultural systems**. Cambridge University Press. p.538, 1992.

MADDONNI, G. A.; OTEGUI, M. E. Intra-specific competition in maize: Contribution of extreme plant hierarchies to grain yield, grain yield components and kernel composition. **Field Crops Research**. v.97, p.155-166, 2006.

MADDONNI, G. A.; OTEGUI, M. E.; CIRILO, A. G. Plant population density, row spacing and hybrid effects on maize canopy architecture and light attenuation. **Field Crops Reseach**. v.71, p.183-193, 2001.

MADDONNI, G. A.; OTEGUI, M. E. Leaf area, light interception, and crop development in maize. **Field Crops research**. v.48, p.81-87, 1996.

MARCONDES, T. Dairy farming in Santa Catarina - Current situation and prospects. **Revista Agropecuària Catarinense**, v.17, n.13, p.20-23, 2004.

MARCHÂO, R. L.; BRASIL, E. M.; XIMENES, P. A. Interception of photosynthetically active radiation and grain yield of densified maize. **Revista Brasileira de Milho e Sorgo.** v.5, n.2, p.170-181, 2006.

MASSIGNAM, A. M. **Quantifying Nitrogen Effects on Crop Growth Processes in Maize and Sunflower**, 2003. 233p. Thesis (PhD) - The University of Queesland, Australian, Brisbane, 2003.

MENDES, R. M. de S.; TAVORA, F. J. A. F.; PINHO, J. L. N.; PITOMBEIRA, J. B. Changes in the source-drain relationship in string beans subjected to different plant densities. **Revista Ciência Agronômica**. v.36, n.1, p.82-90, 2005.

MODARRES, A. M.; HAMILTON, M. D.; DWYER, L. M.; STEWART, D. W.; MATHER, D. E.; SMITH, D. L. Plant population density effectson maize imbreds lines grown in short-season enviroments. **Crop Science**, v.38, p.104-108, 1998.

MINISTRY OF AGRICULTURE AND SUPPLY. **Minimum descriptors for maize** (*Zea mays* L.). Cultivar protection system. Published in the official journal of the union. 13p, 1997.

NORWOOD, C. A. Dryland corn in Western Kansas: Effects of hybrids maturity, planting date and plant population. **Agronomy journal**. v.93, p.540-547, 2001.

OTEGUI, M.; SLAFER, G. A. Increasing cereal yield potential by modifying developmental traits. **Crop Science Congress.** Brisbane, Australia. 2004.

OTEGUI, M. E. Kernel set and flower synchrony within the ear of maize: II. Plant population effects. **Crop Science**. v. 37, p. 448-455. 1997.

ODUM, E. P. **Ecology**. Ed. Guanabara. 434 p. 1986.

PAGANO, E.; MADDONNI, G. A. Intra-specific competition in maize: Early established hierarchies differ in plant growth and biomass partitioning to the ear around silking. **Field Crops Research**. v.15, p.306-320, 2007.

PALHARES, M. **Plant distribution, plant population and corn grain yield**. 2003. 90p. Dissertation (Master's degree in agronomy) - University of São Paulo, Piracicaba, 2003.

PANDOLFO, C.; BRAGA, H. J.; SILVA JÛNIOR, V . P.; MASSIGNAM, A M.; PEREIRA, E. S.; THOMÉ, V. M. R. **Atlas climâticos digital do Estado de Santa Catarina** (CR - Rom). Epagri. Florianópolis, 2002.

RITCHIE, W. S.; HANWAY, J. J.; BENSON, G. O. How the Maize Plant Develops. **Potafos**. n.15, p.1-20, 2003.

ROMANO, M. R. **Physiological performance of the corn crop with plants**

of contrasting architecture: Parameters for growth models. 2005. 100p. Thesis (Doctorate in agronomy) Universidade de Sâo Paulo, Piracicaba, 2005.

SANGOI, L.; GUIDOLIN, A. F.; COIMBRA, J. L. M.; SILVA, P. R. F. Response of maize hybrids grown at different times to plant population and depotting. **Ciência Rural**. v.36, p.1367-1373, 2006a.

SANGOI, L.; SILVA, P. R. F.; SILVA, A. L. da.; ERNANI, P. R.; HORN, D.; STRIEDER, M. L.; SCHIMITT, A.; SCHWEITZER, C. Agronomic performance of maize cultivars in four management systems. **Revista Brasileira de Milho e Sorgo**. v.5, n.2, p.218-231, 2006b.

SANGOI, L.; ARGENTA, G.; SILVA, P. R. F.; MINETTO, T. J.; BISOTTO, V. Management levels in corn cultivation in two contrasting environments: technical-economic analysis. **Ciência Rural**. v.33, p.1021-1029, 2003.

SANGOI, L.; ALMEIDA, M. L.; SILVA, P. R. F.; ARGENTA G. Morphophysiological bases for greater tolerance of modern maize hybrids to high plant densities. **Bragantia**. v. 61, n.2, p.101-110, 2002a.

SANGOI, L.; ALMEIDA, M. L.; GRACIETTI, M. A.; BIANCHET, P.; HORN, D. Maize stalk sustainability in maize hybrids from different eras as affected by plant density. **Revista de Ciências Agroveterinârias**. 2002b.

SANGOI, L.; GRACIETTI, M. A.; RAMPAZZO, C.; BIANCHETTI, P. Response of brazilian maize hybrids from different eras to changes in plant density. **Field Crops Research**. v.79, p. 39-51, 2002c

SANGOI, L.; LECH, V.A.; RAMPAZZO, C.; GRACIETTI, L.C. Dry matter accumulation in maize hybrids under different source-drain relationships. **Pesquisa Agropecuâria Brasileira**. v.37, n.3, p.259267, 2002d.

SANGOI, L.; ALMEIDA, M. L.; LECH, V. A.; GRACIETTI, L. C. Performance of maize hybrids with contrasting cycles as a function of defoliation and plant population. **Scientia Agricola**. v.58, n.2, p.271-276, 2001a.

SANGOI, L.; ALMEIDA, M. L.; RAMPAZZO, C.; GRACIETTI, L. C. Response of maize hybrids grown at different times to increased plant density. In. Santa Catarina Technical Meeting on Corn and Beans. Abstract Chapegó, Epagri, 2001b, p.48-52.

SANGOI, L. Understanding plant density effects on mayze growth and development: An Important issue to maximize grain yield. **Ciência Rural**. v.31, n.1, p.159-168, 2000.

SANGOI, L., SALVADOR, R. J. Influence of plant height and of leaf number on maize production at high plant densities. **Pesquisa Agropecuària Brasileira**. v.33, n.03, p. 297-306. 1998.

SANGOI, L., SALVADOR, R. J. Influence of plant height and leaf number on maize production at high plant densities. **Pesquisa Agropecuària Brasileira**, v.33, n.3, p.297-306, 1998.

SANGOI, L.; SALVADOR, R.J. Agronomic performace of male-sterile and fertile maize genotypes at two plant populations. **Ciência Rural**. v.26, n.3, p.377-388, 1996.

SILVA, P. R. F.; SANGOI, L.; ARGENTA, G.; STRIEDER, M. L. **Plant arrangement and its importance in defining corn productivity**. Porto Alegre, 64p. 2006.

SILVA, P. R. F.; ARGENTA, G.; REZERA, F. Responses of an irrigated hybrid to plant density at three sowing times. **Pesquisa Agropecuària Brasileira**. v.34, p.585-592, 1999.

SINCLAIR, T. R. Historical changes in harvest index and crop nitrogen accumulation. **Crop Science**. v.38, n.3, p.638-643, 1998.

ANNUAL SUMMARY OF THE AGRICULTURE OF SANTA CATARINA 20032004. Florianópolis: Epagri/CEPA, 2005. 146p.

SOUZA, S. N. de. Maize for silage: Agronomic considerations. **Agropecuâria Catarinense**, Florianópolis, v.2, n.2, p.11-14, jun., 1989.

SOARES, A. C. Rescue and conservation. In: SOARES, A. C., MACHADO, A. T., VON DER WEID, J. M., **Milho crioulo, conservação e uso da biodiversidade**. Riode Janeiro: ASPTA, 185p. 1998.

STEINMACHER, N. C. **Physico-chemical characterization, rheological properties and proteins of Creole maize**. 2005. 128p. Florianópolis, Dissertaçâo (Mestrado CAL) - Federal University of Santa Catarina, Florianópolis, 2005.

TAIZ, L.; ZEIGER, E. **Plant physiology**. Porto Alegre, 719p. 2004

TESTA, M. V.; NADAL, R.; MIOR, L. C.; BALDISSERA, I. T.; CORTINA, N. O. **Desenvolvimento Sustentâvel do Oeste Catarinense**. Florianópolis, 247p. 1996.

TOKATLIDIS, L. S.; KOUTROUBAS, S. D. A review of maize hybrids dependence on high plant population and its implications for crop yield stability. **Field Crops Research**. v.88, p.103-114, 2004.

TOKATLIDIS, L. S. The effect of improved potential yield per plant on crop yield potential and optimum plant density in maize hybrids.

Journal of Agricultural Science. v.137, p. 299-305, 2001.

TOLLENAAR, M.; LEE, E. A. Yield potential, yield stability and stress tolerance in maize. **Field Crops Research**. v.75, p.161-169, 2002.

TOLLENAAR, M.; YING, J.; DUVICK, D.N. Genetic gain in corn hybrids from the Northern and Central Corn Belt. **55th Annual Corn and Sorghum Seed Research Conference Proceedings**. Chicago. p.53-62, 2000.

TOLLENAAR, M.; WU, J. Yield improvement in temperate maize is attributable to greater stress tolerance. **Crop Science**. v.39, p.15971604, 1999.

TOLLENAAR, M., AGUILERA, A., NISSANKA, S.P. Grain yield is reduced more by weed interference in an old than in a new maize hybrid. **Agronomy Journal**. v.89, n.2, p.239-246, 1997.

TOLLENAAR, M.; DWYER, L. M.; STWART, D. W. Ear and kernel formation in maize hybrids representing three decades of grain yield improvement in Ontario. **Crop Science**. v.32, p.432-438, 1992.

TOLLENAAR, M.; BRUULSEMA, T. W. Efficiency of maize dry matter production during periods of complete leaf area expansion. **Agronomy Journal**. v.80, p.580-585, 1988.

TSUMANUMA, G. M. **Performance of corn intercropped with different species of brachiaria, in Piracicaba SP.** 2004. 83p. Dissertation (Master's degree in agronomy) - University of São Paulo, Piracicaba, 2004.

WIDDICOMBE, W. D.; THELEN, K. D. Row width and plant density effects on corn forage hybrids. **Agronomy Journal**. v.94, p.326-330, 2002.

WESTGATE, M. E., FORCELLA, F. REICOSK, D. C., SOMSEN, J. Rapid canopy closure for maize production in the northern US Corn Belt: Radiation-use efficiency and grain yield. **Field Crops Research**. v.49, p.249-258, 1997.

UHART, S. A.; ANDRADE, F. H. Nitrogen and carbon accumulation and remobilization during grain filling in maize under different source/sink ratios. **Crop Science**. v.35, p.183-190, 1995.

VOGT, G. A. The **dynamics of the use and management of local maize varieties on family farms**. 2005. 102p. Dissertation (Master's Degree in Plant Genetic Resources) - Federal University of Santa Catarina, Florianópolis, 2005.

8. Appendices

Number of days between emergence and male and female flowering and interval between flowering at plant densities and analysis of variance for the MPA1 maize variety.

	Plant density (Plants ha)$^{-1}$					Analysis of	
Flowering (DAE)	18.000	34.000	47.000	56.000	75.000	variance	CV%
Male	79 b	81 ab	82 ab	83 a	83 a	*	1,8
Female	82 a	83 A	86 a	90 b	92 b	***	2,7
Interval	3 ab	2B	4 ab	7 ac	9 c	*	43,2

Averages followed by the same letter on the same line are not statistically different from each other using the Duncan 5% test (NS not significant, *p<5%, **p<1%, ***p<0.01%).

Appendix 2. Number of days between female flowering (FF) and physiological maturity (MF) at plant densities and analysis of variance for the MPA1 corn variety.

	Plant density (Plants ha- i)					Analysis of variance	CV%
	18.000	34.000	47.000	56.000	75.000		
Interval between FF and MF	75 a	78 a	76 a	66 b	68 b	**	6,1

Averages followed by the same letter on the same line do not differ statistically from each other using the Duncan 5% test (NS not significant, *p<5%, **p<1%, ***p<0.01%).

Appendix 3. Plant height (cm) of plant densities on different days after emergence and analysis of variance for the MPA1 corn variety.

	Plant density (Plants ha- 1)					
DAE	18.000	34.000	47.000	56.000	75.000	Analysis of variance
0	5,69	5,82	5,58	6,08	6,28	NS
4	10,67	10,98	10,79	9,81	10,31	NS
11	16,69	18,28	19,84	19,00	20,00	NS
18	26,66	32,53	27,66	28,94	28,56	NS
25	36,25	44,63	44,09	40,25	39,54	NS
31	47,42	51,72	50,16	47,72	47,63	NS
38	60,91	68,50	67,69	58,00	63,22	NS
46	96,78	105,50	107,44	86,13	101,06	NS
53	134,44	140,81	146,44	118,31	139,31	NS
60	172,75	176,56	182,00	145,88	161,38	NS
67	197,25	198,19	205,28	166,31	181,00	NS
74	241,66	236,13	238,44	203,63	211,94	NS
81	252,44	244,06	249,06	217,69	232,63	NS
88	249,56	243,75	244,13	216,50	231,50	NS
95	248,44	246,19	246,31	221,56	231,31	NS
109	247,56	243,19	245,50	220,25	229,81	NS
123	249,94	244,44	246,38	219,56	227,56	NS
141	246,69	242,56	246,44	214,31	227,63	NS
155	245,19	239,19	243,13	212,81	225,38	NS
197	251,25	250,69	251,81	227,81	229,88	NS

NS. not significant

Appendix 4. Ear insertion height (cm) at plant densities at 197 days after emergence and analysis of variance for maize variety MPA1.

Plant density (Plants ha)$^{-1}$					
18.000	34.000	47.000	56.000	75.000	Analysis of variance
126,19	126,20	138,50	117,47	134,47	NS

NS. not significant

Average stalk diameter (mm) at different plant densities on different days after emergence and analysis of variance for the MPA1 maize variety.

	Plant density (Plants ha⁻ i)											
DAE	18.000		34.000		47.000		56.000		75.000		Analysis of variance	CV%
1	0,00	-	0,00	-	0,00	-	0,00	-	0,00	-	NS	
4	3,29	-	3,71	-	3,17	-	3,27	-	3,17	-	NS	
11	6,02	-	5,88	-	6,33	-	6,18	-	6,07	-	NS	
18	8,84	-	8,89	-	8,19	-	9,32	-	8,08	-	NS	
25	14,61	-	15,20	-	14,90	-	12,96	-	12,21	-	NS	
31	19,14	-	18,83	-	18,47	-	16,49	-	16,10	-	NS	
38	27,3	a	25,13	ab	24,14	abc	21,73	bc	20,78	c	**	9,7
46	32,67	c	28,94	a	28,33	a	24,08	b	23,62	b	***	7,9
53	35,83	c	29,87	a	29,39	a	25,48	b	24,62	b	***	8,8
60	33,05	c	27,64	a	26,46	ab	23,29	ab	22,09	b	***	10,3
67	34,21	d	28,35	a	27,34	ab	23,88	bc	22,17	c	***	9,3
74	33,19	c	27,12	a	26,31	a	23,25	ab	21,63	b	***	9,3
81	32,80	c	26,43	a	25,94	a	21,78	b	20,79	b	***	9,4
88	32,24	c	25,99	a	25,36	a	21,39	b	20,87	b	***	9,6
95	31,87	d	26,15	a	25,14	ab	21,40	bc	20,93	c	***	9,6
109	31,88	c	26,07	a	25,30	a	21,42	b	20,78	b	***	9,7
123	31,49	c	25,69	a	25,51	a	21,15	b	21,05	b	***	9,9
141	31,76	c	25,68	a	25,13	a	21,13	b	20,41	b	***	10,0
155	31,	03c	24.14 ab		24,	79a	22.20 ab		20,32 b		***	10,4
197	31,	39c	25,53 a		24.25 ab		21.40 ab		20,53 b		***	10,9

Averages followed by the same letter on the same line are not statistically different from each other using the Duncan 5% test (NS not significant, *p<5%, **p<1%, ***p<0.01%).

Number of expanded leaves for plant densities at different days after emergence and analysis of variance for the MPA1 maize variety.

	Plant density (Plants ha)$^{-1}$							
DAE	18.000		34.000	47.000	56.000	75.000	Analysis of variance	CV%
1	0,00	-	0,00	0,00	0,00	0,00	NS	
4	3,38	-	3,50	3,31	3,25	3,31	NS	
11	5,06	-	5,00	5,00	5,19	5,31	NS	

18	6,88	-	6,75		6,56		7,00		7,06		NS	
25	8,38	-	8,56		8,75		8,06		7,88		NS	
31	9,87	ab	9,87	ab	10,31	a	9,12	c	9,37	bc	*	3,2
38	12,56	ab	12,75	a	12,81	a	11,50	c	11,93	b	**	3,5
46	15,37	a	15,50	a	15,68	a	14,00	b	13,87	b	***	3,4
53	17,68	a	17,50	a	18,12	a	15,75	b	15,75	b	***	4,1
60	19,12	a	18,93	a	19,43	a	16,81	b	16,81	b	***	4,4
67	20,43	a	20,25	a	20,75	a	18,18	b	18,31	b	***	3,6
74	21,31	a	21,18	a	21,87	a	19,50	b	19,37	b	***	3,0
81	22,06	a	21,87	a	22,25	a	20,31	b	20,50	b	**	3,5
88	22,25	a	22,06	a	22,50	a	20,50	b	20,93	b	**	3,0
95	22,25	a	22,06	ab	22,50	a	20,56	c	21,12	bc	**	2,8
109	22,25	a	22,06	ab	22,50	a	20,56	c	21,12	bc	**	2,9
123	22,25	a	22,06	ab	22,50	a	20,56	c	21,12	bc	**	2,8
141	22,25	a	22,06	ab	22,50	a	20,56	c	21,12	bc	**	2,8
155	22,25	a	22,06	ab	22,50	a	20,56	c	21,12	bc	**	2,9
197	22,25	a	22,06	ab	22,50	a	20,56	c	21,12	bc	**	2,9

Averages followed by the same letter on the same line are not statistically different from each other using the Duncan 5% test (NS not significant, *p<5%, **p<1%, ***p<0.01%).

Appendix 7 - Leaf angle of the first leaf above the main ear and of the 12th leaf, for plant densities at 81 DAE and analysis of variance for maize variety MPA1.

	Plant density (Plants ha- i)					
Leaf	18.000	34.000	47.000	56.000	75.000	Analysis of variance
1st up the spike	24,3	24,1	22,3	23,7	21,9	NS
12°	20,92	19,64	17,10	19,69	16,11	NS

NS. not significant

Appendix 8. Individual leaf area (cm^2) from the base to the top of the vegetative canopy at plant densities and analysis of variance for the MPA1 corn variety.

	Plant density (Plants ha)$^{-1}$						
Leaves	18.000	34.000	47.000	56.000	75.000	Analysis of variance	CV%
1	10,01	10,93	7,45	8,61	8,40	NS	
2	19,10	20,44	15,70	19,27	19,84	NS	
3	35,80	41,11	34,07	36,74	41,51	NS	
4	62,20	72,44	64,90	70,83	76,09	NS	
5	104,78	118,64	105,76	112,33	108,79	NS	
6	143,42	153,52	151,48	154,57	148,63	NS	
7	192,67	191,12	184,83	196,30	182,80	NS	
8	257,99	258,24	237,66	243,16	254,98	NS	
9	347,40	324,81	309,70	324,41	346,38	NS	
10	453,21	413,45	409,78	431,37	444,94	NS	
11	594,06	520,89	534,06	548,59	580,09	NS	
12	727,78	639,78	659,97	620,84	676,58	NS	
13	819,52	729,16	751,16	703,92	752,73	NS	
14	877,12	777,13	791,82	700,26	732,26	NS	

15	899,35	807,20	811,23	674,77	696,11	NS	
16	859,75 a	781,81 a	791,87 a	617,08 b	613,61 b	**	12,6
17	806,50 a	720,10 a	743,55 a	542,29 b	522,65 b	**	12,4
18	750,37 a	649,73 a	658,80 a	454,38 b	442,77 b	***	12,6
19	637,38 a	546,67 a	564,16 a	428,34 b	368,82 b	***	13,4
20	562,00 a	450,50 b	468.74 ab	313,85 c	292,25 c	***	16,9
21	400,28	373,64	386,33	297,36	251,78	NS	
22	406,88	286,10	302,24	251,78	216,09	NS	
23	378,31	239,70	463,43	205,43	172,50	NS	
24	288,80	199,65	374,39		112,20	-	
25	270,49		243,15			-	
26	154,05					-	

Averages followed by the same letter on the same line do not differ statistically from each other using the Duncan 5% test (NS not significant, *p<5%, **p<1%, ***p<0.01%).

Appendix 9. Total leaf area per plant (cm^2) at different DAE for the plant densities and analysis of variance for the MPA1 corn variety.

	Plant density (Plants ha)$^{-1}$					Analysis of	
DAE	18.000	34.000	47.000	56.000	75.000	variance	CV%
4	84,00	104,42	84,95	82,89	87,63	NS	
11	188,59	223,94	190,39	226,90	239,84	NS	
18	386,60	420,56	362,96	436,69	416,01	NS	
25	529,19	541,43	601,90	504,84	508,48	NS	
31	857,09	889,97	886,12	710,20	761,25	NS	
38	1615,14	1598,70	1661,55	1353,96	1467,10	NS	
46	2676,42	2698,14	2756,54	2000,61	2270,69	NS	
53	4174.86 ab	4055.45 ab	4266,77 a	3007,06 c	3287.49 bc	*	15,3
60	5609,56 a	5181.18 ab	5503,59 a	3786,08 c	4089.50 bc	*	17,6
67	7229,20 a	6655.58 abc	7121.97 ab	5064,87 c	5483.27 bc	*	16,1
74	9233,75 a	8290.58 ab	8729,86 a	6642,57 b	6734,52 b	*	14,9
81	9888,91 b	8800.23 ab	9064.21 ab	7215,74 a	7362,69 a	*	14,3
88	9964,66 b	8860.69 ab	9131.66 ab	7300,51 a	7501,63 a	*	13,8
95	9974,88 b	8860.69 ab	9131.66 ab	7307,21 a	7532,63 a	*	13,7
104	9974,88 b	8860.69 ab	9131.66 ab	7307,21 a	7532,63 a	*	13,7
109	9974,88 b	8860.69 ab	9131.66 ab	7307,21 a	7532,63 a	*	13,7
116	9974,88 b	8860.69 ab	9131.66 ab	7307,21 a	7532,63 a	*	13,7
123	9974,88 b	8860.69 ab	9131.66 ab	7307,21 a	7532,63 a	*	13,7
134	9974,88 b	8860.69 ab	9131.66 ab	7307,21 a	7532,63 a	*	13,7
141	9974,88 b	8860.69 ab	9131.66 ab	7307,21 a	7532,63 a	*	13,7
148	9974,88 b	8860.69 ab	9131.66 ab	7307,21 a	7532,63 a	*	13,7
156	9974,88 b	8860.69 ab	9131.66 ab	7307,21 a	7532,63 a	*	13,7
161	9974,88 b	8860.69 ab	9131.66 ab	7307,21 a	7532,63 a	*	13,7
170	9974,88 b	8860.69 ab	9131.66 ab	7307,21 a	7532,63 a	*	13,7

Averages followed by the same letter on the same line are not statistically different from each other using the Duncan 5% test (NS not significant, *p<5%, **p<1%, ***p<0.01%).

Appendix 10 - Green area per plant (cm2) for different plant density DAE and analysis of variance for maize variety MPA1.

DAE	Plant density (Plants ha)$^{-1}$	CV%

	18.000		34.000		47.000		56.000		75.000		Analysis of variance	
4	84,00		104,42		84,95		82,89		87,63		NS	
11	188,59		223,94		190,39		226,90		239,84		NS	
18	386,60		420,56		362,96		436,69		416,01		NS	
25	529,19		541,43		601,90		504,84		508,48		NS	
31	849,55		873,07		873,71		692,94		751,63		NS	
38	1586,46		1568,12		1634,89		1324,47		1443,53		NS	
46	2606,32		2648,62		2708,44		1953,46		2210,91		NS	
53	4025.80	ab	3950.05	ab	4150,67	a	2916,82	c	3167.67	bc	*	15,8
60	5316,14	a	4891,61	ab	5219,46	a	3520,69	c	3745,18	bc	*	17,9
67	6805,64	a	6265,09	ab	6665,16	a	4691,79	b	4997,36	b	*	16,8
74	8753,59	a	7841,94	ab	8192,94	a	6200,02	b	6194,35	b	*	15,0
81	9346,53	b	8195,50	ab	8371,49	ab	6598,27	a	6611,92	a	*	14,5
88	9236,74	a	7957,06	a	8017,85	a	6303,97	b	6215,70	b	**	14,1
95	9002,08	a	7655,02	a	7503,89	ab	5968,35	b	5928,27	b	**	14,2
104	8874,12	a	7590,63	a	7432,13	a	5802,95	b	5742,57	b	**	14,3
109	8789,99	a	7517,15	a	7274,27	ab	5725,85	b	5811,22	b	**	15,0
116	8686,89	a	7328,13	ab	7082,24	abc	5691,61	bc	5554,87	c	**	15,3
123	8161,03	c	6454,59	a	6091,51	ab	4898,70	ab	4670,51	b	**	16,8
134	6475,82	b	5302,88	ab	4965,11	ab	3906,87	a	3629,35	a	*	24,0
141	5107,73		4071,88		3783,88		3164,09		2295,73		NS	
148	3317,96		3030,30		2026,24		2112,82		1259,93		NS	
156	534,44		338,72		253,57		466,61		251,96		NS	
161	10,50		0,00		64,91		0,00		172,82		NS	
170	0,00		0,00		0,00		0,00		0,00		NS	

Averages followed by the same letter on the same line are not statistically different from each other using the Duncan 5% test (NS not significant, *p<5%, **p<1%, ***p<0.01%).

Appendix 11 - Relative senescent leaf area per plant (cm^2) for different plant densities and analysis of variance for the MPA1 maize variety.

	Plant density (Plants ha)$^{-1}$					Analysis of	
DAE	18.000	34.000	47.000	56.000	75.000	variance	CV%
4	0,000	0,000	0,000	0,000	0,000	-	
11	0,000	0,000	0,000	0,000	0,000	-	
18	0,000	0,000	0,000	0,000	0,000	-	
25	0,000	0,000	0,000	0,000	0,000	-	
31	0,001	0,002	0,001	0,002	0,001	NS	
38	0,003	0,003	0,003	0,004	0,003	NS	
46	0,007	0,006	0,005	0,006	0,008	NS	
53	0,015	0,012	0,013	0,012	0,016	NS	
60	0,029	0,033	0,031	0,036	0,046	NS	
67	0,042 a	0,044 a	0,050 a	0, 051a	0,064 b	**	14,5
74	0,048 a	0,050 a	0,058 a	0,060 a	0, 07 1b	**	17,0
81	0,054 b	0.068 ab	0.075 ab	0.084 ac	0,099 c	*	19,5
88	0,072 c	0. 101a c	0,121 a	0.136 ab	0,170 b	**	22,1
95	0,097 c	0.136 bc	0.178 ab	0.183 ab	0,212 a	**	20,8
104	0,110 c	0.140 bc	0.180 ab	0,200 a	0,230 a	**	19,2

109	0,110 b	0,150 b	0,200 a	0,210 a	0,220 a	***	16,6
116	0,120 c	0.170 bc	0.220 ab	0.220 ab	0,260 a	**	20,5
123	0,180 b	0.270 ab	0,330 a	0,320 a	0,370 a	*	24,0
134	0,351	0,402	0,456	0,465	0,518	NS	
141	0,488	0,540	0,586	0,567	0,695	NS	
148	0,667	0,658	0,778	0,711	0,833	NS	
156	0,946	0,962	0,972	0,936	0,967	NS	
161	0,999	1,0	0,993	1,0	0,977	NS	
170	1,0	1,0	1,0	1,0	1,0	NS	

Averages followed by the same letter on the same line are not statistically different from each other using the Duncan 5% test (NS not significant, *p<5%, **p<1%, ***p<0.01%).

Appendix 12. Leaf area index for different DAE of plant densities and analysis of variance for the MPA1 corn variety.

	Plant density (Plants ha)$^{-1}$										Analysis of	
DAE	18.000		34.000		47.000		56.000		75.000		variance	CV%
4	0,01	c	0,04	ac	0,05	a	0,06	ab	0,08	b	**	36,3
11	0,03	b	0,08	ab	0,11	a	0,18	d	0,23	c	***	24,7
18	0,07	c	0,16	b	0,21	b	0,34	a	0,41	a	***	19,5
25	0,10	d	0,21	c	0,36	b	0,40	ab	0,50	a	***	22,1
31	0,16	d	0,34	c	0,52	a	0,55	a	0,75	b	***	21,2
38	0,31	d	0,62	c	0,98	a	1,05	a	1,44	b	***	17,2
46	0,52	d	1,05	c	1,62	a	1,56	a	2,21	b	***	16,6
53	0,80	d	1,58	c	2,49	a	2,33	a	3,16	b	***	16,7
60	1,06	d	1,95	c	3,13	ab	2,81	b	3,74	a	***	19,1
67	1,36	d	2,50	c	3,99	a	3,75	a	4,99	b	***	17,3
74	1,75	d	3,13	c	4,91	a	4,96	a	6,19	b	***	14,2
81	1,86	d	3,27	c	5,02	a	5,27	a	6,61	b	**	12,8
88	1,84	d	3,18	c	4,81	a	5,04	a	6,21	b	***	12,0
95	1,80	d	3,06	c	4,50	a	4,77	a	5,92	b	***	12,0
104	1,77	d	3,03	c	4,45	a	4,64	a	5,74	b	***	12,5
109	1,75	d	3,00	c	4,36	a	4,58	a	5,81	b	***	11,0
116	1,73	d	2,93	c	4,24	a	4,55	a	5,55	b	***	13,5
123	1,63	d	2,58	c	3,65	b	3,91	ab	4,67	a	***	17,1
134	1,29	c	2,12	bc	2,97	ab	3,12	ab	3,62	a	**	30,1
141	1,02		1,63		2,27		2,53		2,30		NS	
148	0,66		1,21		1,22		1,69		1,26		NS	
156	0,11		0,14		0,15		0,37		0,25		NS	
161	0,0		0,0		0,04		0,0		0,17		-	
170	0,0		0,0		0,0		0,0		0,0		-	
197	0,0		0,0		0,0		0,0		0,0		-	

Averages followed by the same letter on the same line are not statistically different from each other using the Duncan 5% test (NS not significant, *p<5%, **p<1%, ***p<0.01%).

Appendix 13 - Total phytomass (kg ha^{-1}) for different DAE at different plant densities and analysis of variance for the MPA1 maize variety.

	Plant density (Plants ha)$^{-1}$										Analysis of	
DAE	18.000		34.000		47.000		56.000		75.000		variance	CV%
1	0,41	c	0,84	bc	1,39	a	1,24	ab	1,72	a	***	28,4
11	6,04	d	14,89	c	24,49	a	27,38	a	38,27	b	***	11,3
18	24,00	d	52,91	a	63,03	a	88,53	c	112,77	b	***	21,2

49	600,87	b	1986,46	bc	4520,92	c	7083,85	a	7738,70	a	***	28,8
84	5065,16	b	6011,59	ab	7910,57	a	7695,25	a	10621,8	c	***	18,3
112	7357,00	c	10427,3	bc	14030,6	ab	16434,2	a	15579,0	a	**	24,3
154	11861,8	-	14045,1	-	15645,3	-	14819,7	-	14453,2	-	NS	
197	7045,71	b	11636,63	a	11491,31	a	11135,66	a	11821,26	a	*	16,5

Averages followed by the same letter on the same line do not differ statistically from each other using the Duncan 5% test. (NS not significant, *p<5%, **p<1%,

Appendix 14 - Stem phytomass (kg ha^{-1}) for different DAE at different plant densities and analysis of variance for the MPA1 maize variety.

	Plant density (Plants ha)$^{-1}$										Analysis of	
DAE	18.000		34.000		47.000		56.000		75.000		variance	CV%
11	2,58	d	6,21	c	10,5	a	11,54	a	16,25	b	***	13,3
18	11,18	b	25,37	ab	28,84	a	39,00	a	53,92	c	***	31,3
49	284,58	b	940,55	b	2056,53	c	3294,65	a	3540,51	a	***	33,2
84	2773,13	b	3255,45	b	4507,91	a	4630,93	a	6316,75	c	***	18,0
112	2762,06	c	4014,04	bc	5315,64	ab	6566,71	a	6313,37	ab	*	29,5
154	3373,98		4280,78		5004,51		5505,96		4779,97		NS	
197	2371,53		3425,26		3171,34		3857,49		3453,16		NS	

Averages followed by the same letter on the same line are not statistically different from each other using the Duncan 5% test (NS not significant, *p<5%, **p<1%, ***p<0.01%).

Appendix 15 - Proportion of stems per plant for different DAE at different plant densities and analysis of variance for the MPA1 maize variety.

	Plant density (Plants ha)$^{-1}$					Analysis of
DAE	18.000	34.000	47.000	56.000	75.000	variance
11	0,43	0,42	0,43	0,42	0,42	NS
18	0,46	0,47	0,46	0,43	0,46	NS
49	0,47	0,47	0,45	0,46	0,45	NS
84	0,55	0,54	0,57	0,60	0,60	NS
112	0,38	0,39	0,40	0,40	0,39	NS
154	0,29	0,30	0,32	0,39	0,33	NS
197	0,34	0,30	0,28	0,35	0,29	NS

NS. not significant

Appendix 16: Green leaf phytomass (kg ha^{-1}) in relation to total phytomass for different DAE at different plant densities and analysis of variance for the MPA1 maize variety.

	Plant density (Plants ha)$^{-1}$										Analysis of	
DAE	18.000		34.000		47.000		56.000		75.000		variance	CV%
11	3,46	d	8,68	c	13,99	a	15,84	a	22,02	b	***	11,3
18	12,81	d	27,54	a	34,18	a	49,53	c	58,85	b	***	15,2
49	316,28	b	1045,91	b	2464,39	c	3789,20	a	4198,19	a	***	25,8
84	1097,50	b	1496,50	b	2017,05	a	2067,76	a	2663,83	c	***	25,4
112	1159,58	c	1706,72	bc	2107,76	ab	2596,92	a	2594,09	a	**	15,4
154	88,17	b	511,29	a	621,40	a	409,61	a	311,69	ab	*	22,7
197	0		0		0		0		0		-	

Averages followed by the same letter on the same line are not statistically different from each other using the Duncan 5% test (NS not significant, *p<5%, **p<1%, ***p<0.01%).

Appendix 17 - Proportion of green leaves in relation to total phytomass for plant densities and analysis of variance for maize variety MPA1.

DAE	Plant density (Plants ha)$^{-1}$					Analysis of variance
	18.000	34.000	47.000	56.000	75.000	
11	0,57	0,58	0,57	0,58	0,58	NS
18	0,54	0,53	0,54	0,57	0,54	NS
49	0,53	0,53	0,55	0,54	0,55	NS
84	0,22	0,25	0,26	0,27	0,25	NS
112	0,16	0,16	0,16	0,16	0,16	NS
154	0,01	0,04	0,04	0,03	0,02	NS
197	0	0	0	0	0	-

NS. not significant

Appendix 18: Phytomass of senescent leaves (kg ha^{-1}) for different DAE at different plant densities and analysis of variance for the MPA1 maize variety.

DAE	Plant density (Plants ha)$^{-1}$					Analysis of variance	CV%
	18.000	34.000	47.000	56.000	75.000		
49	0	0	0	0	0	NS	
84	10,01 a	50,19a	94,95a	99,22a	258,46 b	**	79,2
112	30,67 d	152.89 cd	284.56 bc	517.76 ab	474,30 a	***	43,3
154	1130,04	1161,38	1398,58	1631,00	1635,93	NS	
197	589,75 b	1036,40 a	1081,83 a	1166,29 a	1257,07 a	*	22,2

Averages followed by the same letter on the same line are not statistically different from each other using the Duncan 5% test (NS not significant, *p<5%, **p<1%, ***p<0.01%).

Appendix 19 - Hanging phytomass (kg ha^{-1}) for different DAE at different plant densities and analysis of variance for the MPA1 maize variety.

DAE	Plant density (Plants ha)$^{-1}$					Analysis of variance	CV%
	18.000	34.000	47.000	56.000	75.000		
49	0	0	0	0	0	NS	
84	17,01 c	25.24 bc	33.48 ab	39,12 a	56,78 d	***	22,7
112	12,78 c	19.63 bc	24.36 ab	28.96 ab	30,55 a	**	25,2
154	7,78	11,52	11,26	12,94	12,27	NS	
197	6,83	7,22	5,64	8,56	8,97	NS	

Averages followed by the same letter on the same line are not statistically different from each other using the Duncan 5% test (NS not significant, *p<5%, **p<1%, ***p<0.01%).

Appendix 20. Grain yield (kg ha^{-1}) (RG); Number of ears plant^{-1} (NEP); Number of ears m^{-2} (NE), Number of grains m^{-2} (NGA) Number of grains ear^{-1} (NGE); 1000 kernel weight (PMG); kernel/ear ratio (RGE) and harvest index (CI) of plant densities (DP) and analysis of variance of the MPA1 corn variety.

Densities	RG	NEP	NE	NGA	NGE	PMG	RGE	IC

(Plants ha)$^{-1}$								
18.000	2821,70 b	1,13 b	2,08 d	821,94 b	408,65	a 299,79	0,60	0,36
34.000	5500,60 a	1,06 b	3,58 a	1514,7 a	421,84	a294 ,40	0,68	0,38
47.000	5735,70 a	0,91 a	4.31 ab	1625,3 a	374,02	ab 283.61	0,71	0,40
56.000	5194,60 a	0,88 a	4,98 a	1530,8 a	305,27	bc 267.14	0,70	0,36
75.000	5748,20 a	0,82 a	6,10 c	1662,8 a	275,46	c273 ,15	0,68	0,39
Analysis of Variance	***	**	***	**	**	NS	NS	NS
CV%	15,2	9,8	13,2	16	14,7	-	-	-

Averages followed by the same letter on the same line are not statistically different from each other using the Duncan 5% test (NS not significant, *p<5%, **p<1%, ***p<0.01%).

Appendix 21: Phytomass of weeds for the densities at 155 DAE and analysis of variance for the MPA1 maize variety.

	Plant density (Plants ha)$^{-1}$					Analysis of	
DAE	18.000	34.000	47.000	56.000	75.000	variance	CV%
155	2238,8 c	1063,8 a	491.3 ab	457.5 ab	405,0 b	***	40,9

Averages followed by the same letter on the same line do not differ statistically from each other using the Duncan 5% test (NS not significant, *p<5%, **p<1%, ***p<0.01%).

Appendix 22. Percentage of lodging and plant breakage at different DAE of the plant densities and analysis of variance for the MPA1 corn variety.

	Plant density (Plants ha)$^{-1}$					Analysis of	
DAE	18.000	34.000	47.000	56.000	75.000	variance	CV%
155	0 c	1.2 bc	4.1 ab	3,8 ab	4,8 a	*	66,8
197	4,2	8,3	9,5	12,6	14,1	NS	

Averages followed by the same letter on the same line do not differ statistically from each other using the Duncan 5% test (NS not significant, *p<5%, **p<1%, ***p<0.01%).

Appendix 23. Correlation analysis between grain yield (GY), phytomass (FM), leaf area index (LAI), harvest index (CI), thousand grain weight (MGW), number of ears per plant (NEP), number of ears per area (NE), number of grains per ear (NGE), number of grains per area (NGA), lodging and plant breakage (AQ), height (ALT), stem diameter (DCOL), number of leaves (NFH) and total leaf area per plant (AFTP).

Correlation analysis	RG	FM	IAF	IC	PMG	NEP	NE	NGE	NGA	AQ	ALT	DCOL	NFH	AFTP
RG	1,0	---	---	---	---	---	---	---	---	---	---	---	---	---
FM	0.87 **	1,0	---	---	---	---	---	---	---	---	---	---	---	---
IAF	0.64 **	0.57 **	1,0	---	---	---	---	---	---	---	---	---	---	---
IC	0.39 ns	-0.05 ns	0.21 ns	1,0	---	---	---	---	---	---	---	---	---	---
PMG	-0.31 ns	-0.27 ns	-0.50 *	-0.04 ns	1,0	---	---	---	---	---	---	---	---	---
NEP	-0.42 ns	-0.30 ns	-0.85 **	-0.34 ns	0.48 *	1,0	---	---	---	---	---	---	---	---
NE	0.72 **	0.73 **	0.89 **	0.06 ns	-0.54 *	-0.66 **	1,0	---	---	---	---	---	---	---

NGE	-0.05 ns	-0.22 ns	-0.59 **	0.41 ns	0.23 ns	0.37 ns	-0.65 **	1,0	---	---	---	---	---	---
NGA	0.96 **	0.83 **	0.68 **	0.41 ns	-0.51 *	-0.49 *	0.75 **	-0.05 ns	1,0	---	---	---	---	---
AQ	0.48 *	0.39 ns	0.46 *	0.22 ns	-0.53 *	-0.47 *	0.48 *	-0.06 ns	0.55 *	1,0	---	---	---	---
ALT	-0.14 ns	-0.14 ns	-0.19 ns	0.08 ns	0.50 *	0.22 ns	-0.41 ns	0.33 ns	-0.22 ns	-0.35 ns	1,0	---	---	---
DCOL	-0.63 **	-0.58 **	-0.71 **	-0.10 ns	0.46 *	0.63 **	-0.82 **	0.55 *	-0.63 **	-0.43 ns	0.52 *	1,0	---	---
NFH	-0.17 ns	-0.18 ns	-0.37 ns	0.18 ns	0.46 *	0.33 ns	-0.52 *	0.56 **	-0.22 ns	-0.28 ns	0.55 *	0.68 **	1,0	---
AFT	-0.33 ns	-0.38 ns	-0.41 ns	0.19 ns	0.50 *	0.38 ns	-0.62 **	0.56 **	-0.36 ns	-0.41 ns	0.73 **	0.87 **	0.79 **	1,0

NS not significant, *p<5%, **p<1%, ***p<0.01%).

Printed by Books on Demand GmbH, Norderstedt / Germany